Guido Küstel

Roasting of Gold and Silver Ores

And the Extraction of Their Respective Metals Without Quicksilver

Guido Küstel

Roasting of Gold and Silver Ores
And the Extraction of Their Respective Metals Without Quicksilver

ISBN/EAN: 9783744678513

Printed in Europe, USA, Canada, Australia, Japan

Cover: Foto ©berggeist007 / pixelio.de

More available books at **www.hansebooks.com**

ROASTING

OF

GOLD AND SILVER ORES

AND THE

EXTRACTION

OF THEIR

RESPECTIVE METALS

WITHOUT QUICKSILVER.

BY G. KUSTEL,

MINING ENGINEER AND METALLURGIST,

Author of "Nevada and California Processes of Silver and Gold Extraction,"
and "Concentration of all Kinds of Ores."

Illustrated with Numerous Engravings.

PUBLISHED AND SOLD BY

DEWEY & CO., PROPRIETORS SCIENTIFIC PRESS.

SAN FRANCISCO, 1870.

PREFACE.

The publication of this Treatise is due solely to the many inquiries concerning the "Leaching, Solving and Precipitation Process for Silver Ores," now successfully practiced in Sonora, Mexico, where it has been lately introduced by Mr. Ottocar Hofmann.

In consideration of the very important preparation of the ore, before it is subjected to the Solving Process,—namely, the Roasting,—I have thought it proper to devote considerable space to the description of different modifications of this operation, which is regulated by the peculiarity of the ore, and by the subsequent treatment. It is impossible to give any one way which will be suitable in every case ; for this reason, and in order to cover all cases as far as possible, a detailed description of different modes of Roasting will not be superfluous.

The Solving Process, as now practiced, is a very economical method for the extraction of silver, for the reason that no quicksilver and no castings are used except what are needed for crushing. Mills in Mexico being dependent on San Francisco for the shoes, dies, gearing, etc., of amalgamating pans, millmen there know how to appreciate a process confined to wooden tubs requiring no power. A comparatively small capital is necessary for building up such works, and hence there is a more reasonable ratio established between the amount of money which must be expended on the works and the real value of the mine, than where other more expensive

machinery is employed,—a circumstance which, being insuf-
ficiently regarded, is often the source of failure.

Mr. O. Hofmann commenced first with the "Chlorination
Process" (§ 82), but finding great difficulty in obtaining the
regular supply of sulphuric acid and manganese from San
Francisco, abandoned the chlorination with cold chlorine gas,
which is indispensable in the presence of gold.   Another difficulty was in obtaining a good article of sulphide of sodium.
He tried to extract the potash from ashes, and to use this in
place of soda, but decided finally in favor of lime, which is
found in abundance.   From this the sulphide of calcium is
easily manufactured on the spot.   Sulphide of calcium was
first applied by Kiss (§ 80).

The Solving Process is very simple, and readily performed
by common workmen ; besides the lime, only brimstone
must be provided, in order to prepare the necessary chem-
icals for solving and precipitation.   It is a general but
erroneous belief, that the solving is a slow process.   An
amalgamating pan is charged with 500 to 1,000 pounds of
roasted ore, and treated at least six hours, and therefore
turns out at most two tons in 24 hours ; while a box or vat
of proper size used in the Solving Process, can work from
four to five tons in the same time.

Only those ores are treated by this process which absolutely require roasting ; which, however, with improved furnaces, is not so expensive as it used to be.   The chloride
ores alone can be leached directly without roasting, and this
when there is no other silver combination in them.

                                        G. Kustel.
March, 1870.

I. INTRODUCTION.

Classification of Ores.

1. Ores may be classified : *a*. According to the metal, the extraction of which is principally remunerative ; as silver ores, lead ores, copper ores, etc. *b*. According to the metallurgical treatment ; as roasting ores, smelting ores, amalgamating ores, etc. *c*. According to the predominant gangue, as calcareous ores, quartzose or ochery ores. *d*. According to the predominant metallic mineral; as sulphuret ores, chloride ores, carbonates, etc.

Important Silver Ores.

2. The most important silver ores are those found in such quantities as to be an object of metallurgical operations. The principal minerals of this kind are the following :

A. Real Silver Ores. a. Sulphuret of Silver, or silver glance, with 87 per cent. of silver. It is of common occurrence, and is the most suitable of the silver sulphurets for pan amalgamation without roasting. *b.* *Brittle Silver Ore*, or sulphuret of silver and antimony. This mineral contains 68 per cent. of silver, and is quite common. *c.* *Polybasite*, sulphuret of silver, antimony and some arsenic, with 75 per cent. of silver. Brittle silver ore and polybasite are both tractable in pans without roasting, although not so readily as the simple sulphuret. All other sulphureted silver ores require roasting. *d.* *Ruby Silver*. The dark red silver ore, or antimonial variety, with 59 per cent., and the light red silver ore, or arsenical variety, with 65 per cent. of silver, are valuable minerals. They occur quite frequently in Nevada, Idaho, Montana, Mexico, etc. *e.* *Miargyrite*, sulphuret of silver and antimony; 36.5 per cent. of silver; Idaho, Montana, etc. *f.* *Stromeyerite*, or silver copper glance, a sulphuret of silver and copper containing up to 53 per cent. of silver; Nevada, Arizona, etc. *g.* *Horn Silver*, or chloride of silver, with 75 per cent. of silver; occurs massive in White Pine, Nevada; prepared by nature for the pan amalgamation. *h.* *Stetefeldtite* and *Partzite*, with up to 25 per cent. of silver, are oxide ores which occur very frequently in Nevada, Arizona, etc.

B. *Argentiferous Ores.* *a.* *Silver-fahl-ore*, argentiferous gray copper ore. It contains silver in very variable proportions up to 31 per cent. This ore is quite common, and for this reason is important. It is also one of the most rebellious ores, containing copper, antimony, arsenic, sulphur, lead, iron, zinc, and sometimes gold and quicksilver. *b. Argentiferous Lead Ores*, galena, or sulphuret of lead, lead glance. Generally, this is not rich in silver, containing from $20 to $60 per ton. Specimens assay sometimes as high as $300. The fine grained variety is generally considered richer than the coarse crystallized kind, but this has not been observed to be the case in Nevada and Arizona. *c. Cerusite*, carbonate of lead. If pure, without admixture of copper and other carbonates, it is poor in silver in most cases. Raw, it amalgamates only too readily in pans. Smelting is the only proper way of treating galena and cerusite. *d. Argentiferous Zincblende.* Sulphuret of zinc. Pure zincblende contains usually only traces of silver; often, however, it assays well, even up to $400 per ton, although no other silver ore can be detected with it. In some mines the argentiferous zincblende prevails, and is the most important ore. It requires a great heat in roasting. *e. Argentiferous Pyrites.* Copper and iron pyrites are poor in silver, but often auriferous. Pyrite is a valuable companion for silver ores which have to be treated by a chloridizing roasting, on account of its amount of sulphur, which is necessary for the decomposition of salt.

Difference between Real Silver Ores and Argentiferous Ores.

3. Real silver ores have mostly an unvariable amount of silver. Real silver minerals admit an approximate estimate of the value of the ore, if the proportion of ore and gangue is considered, without making an assay. With the argentiferous ores it is different. Fahl ore, for instance, may be very poor or very rich,'and its value can be ascertained only by an assay. There are no means of estimating the richness of argentiferous ores "by sight."

Important Combinations.

4. With the exception of a few metal oxides of iron, zinc, tin, manganese, and, among silver ores, of the stetefeldtite, etc., the most important, because most frequent ores, are the sulphureted varieties. Sulphur is the most formidable obstacle to the metallurgist in extracting metals from their respective ores. Desulphurization has been a subject of most diligent and numerous experiments. The oldest method is the application of heat, which is still in use, notwithstanding the many attempts in modern times to dispense entirely with fire or to modify its application so as to perform the process more perfectly and in a shorter time.

Means of Desulphurization.

5. The desulphurization of ores is effected: *a*. By heating with free admission of air. This is the common way of "roasting," and the most important, and is effected either in kilns, heaps, etc., or in reverberatory furnaces. As soon as the sulphureted ore is heated to a certain degree, one part of the sulphur escapes as sulphurous acid; another is converted into sulphuric acid, which is also decomposed by an increased heat. Some sulphurets (iron pyrites) lose their sulphur without the application of heat, being decomposed by exposure to the action of air for a long time. This way is sometimes practiced on gold-bearing pyrites. *b.* By heating with exclusion of air. Only the sulphides of gold and platinum are decomposed perfectly by this method. Other sulphureted ores lose their sulphur only in part, being reduced to a lower state of sulphide. Sulphuret of silver (Ag S) remains undecomposed. Cinnabar, sulphide of antimony (Sb S^3) and sulphide of arsenic volatilize unchanged. Iron pyrites (Fe S^2) gives up 23 per cent. of its sulphur, being reduced to magnetic pyrites, and, by a strong heat, to proto-sulphide of iron (Fe S), not further reducible. Also sulphide of zinc (zincblende), remains undecomposed. Copper glance retains its sulphur, and copper pyrites loses only one part of the sulphur which is combined with the iron in it. Galena (Pb S) is reduced to a lower state (Pb^4 S), a part of the lead separating out in a metallic state. *c.* By superheated steam. Sulphurets not evolving sulphur by the last process, lose their sulphur slowly on the application of steam, sulphureted hydrogen and sulphurous acid being formed. Experiments made by Regnault showed that desulphurization is

1*

effected more perfectly if air is admitted. Roasting in reverberatory furnaces is always effected by the oxygen of the air and by steam, as there is no fuel used which contains less than 25 to 30 per cent. of water. Superheated steam has been tried in different ways on sulphurets with the highest expectations, but with no better results for practical use than are given in the ordinary way by the steam obtained from fuel. It may be useful in many instances to have more steam than is thus obtained, but this increases considerably the expense of roasting; as, for instance, in Patera's application of steam in roasting silver ores, tried principally with the intention of expelling antimony, arsenic, etc. Another application of superheated steam, with exclusion of air, is Hagan's method, which may prove successful on pyritous ores, having at the same time the advantage of being a very cheap method. *d*. By heating with metals, alkalies or alkaline earths, for which the sulphur has a greater affinity. The affinity of sulphur for the following metals decreases in the order in which they stand, being strongest for the first and weakest for the last: Copper, iron, tin, zinc, lead, silver, antimony, arsenic. Each of these metals can be desulphurized by the next preceding, though with difficulty; but more easily by one further off. Practical use of this property is made in smelting galena with the addition of metallic iron or iron ore. Sulphide of silver in crucibles is decomposed by stirring the liquid with red hot iron. Quicksilver is obtained from cinnabar by heating the latter with lime, which takes up the sulphur, etc. *e.* Carbon has no great affinity for sulphur; the use of charcoal for desulphurization of ores is therefore an inferior method. So is also the use of carbonic acid.

Result of Desulphurization.

6. The direct extraction of metals from sulphurets, either by smelting or amalgamation, is not practicable. In smelting, the sulphurets melt very readily, but only a small part, if any, of the metal is obtained, while the greater part runs out combined with the sulphur as matt. For this reason the roasting of sulphureted ore for the purpose of smelting is indispensable, unless iron is added. Such roasting or burning takes often many weeks, or months. The direct amalgamation, also, of sulphurets gives a very poor result, except in the case of silver glance. By means of electricity, combined with the chemical action of sulphate of copper and salt, the silver and gold sulphurets are decomposed; but, with the exception of the patio amalgamation, no process has yet been publicly demonstrated as really practical for the treatment of all kinds of raw sulphurets. The desulphurization is therefore still a most important preparation for the extraction of metals. The general effect of roasting is that the metals are oxidized. Only gold and silver are transformed into a metallic

condition; and of the silver, moreover, a large percentage is always found as a sulphate, even when
the roasting is well performed. Some of the silver
combines as an oxide with antimony and silica, if
present. All the oxides obtained by desulphurization must be again deoxidized in order to get them
in a metallic state.

Means of Reduction or Deoxidation.

7. Heating alone will reduce the oxides of the
precious metals only. Oxide of gold does not
occur in nature, neither is it obtained in any of the
metallurgical processes. Oxide of silver is also
unimportant; it is formed, to a small extent, in cupellation (and taken up by the litharge), in smelting silver ores combined with silica, and in roasting
silver ores in the presence of antimony, arsenic,
etc.

The most powerful agent of reduction is carbon
(charcoal, coke, etc.) and carbonic oxide. In all
smelting in blast furnaces, the carbonic oxide is
the real reducer. The burning coal, under the influence of the compressed air, produces carbonic
acid, melting at the same time the ore; the carbonic acid, passing through the glowing coal above
the melting region, gives up a part of its oxygen to
the coal, and is reduced thereby to carbonic oxide,
which in turn takes up oxygen again, from the
metal oxides, reducing them to a metallic state; a
contact of ore oxides with carbon is therefore not

necessary for the purpose of reduction. All metals do not retain their oxygen with equal tenacity, but some part with it much more easily than others. For instance, lead, copper, bismuth, antimony, cobalt and nickel, require for their reduction a darker or lighter red heat, while iron, zinc and tin are reduced only at a white heat. But also hydro-, gen and carbureted hydrogen, created by the burning fuel, are powerful reducing agents.

Metal oxides in solution are reduced and precipitated in a metallic condition by other metals. On this principle copper is precipitated by metallic iron, which goes into solution in place of the copper; sulphate of silver, in Ziervogel's process, is precipitated by copper, etc. Also, by aid of the electro-galvanic stream, metals are reduced to a metallic state from their solutions.

Desulphurization of Silver Ores not Efficient.

8. Although, by mere desulphurization the silver is to a great extent converted into a metallic state, this is not always its most suitable condition except for smelting. Almost all silver extracted in the United States is obtained by amalgamation, smelting being confined to a few localities where the ore contains such a high percentage of lead that its amalgamation is impossible. It would seem as if metallic silver should amalgamate more easily than if combined with another substance.

This, however, is not the case. The silver, after roasting, is generally coated with the oxides of volatile base metals, which prevent its ready amalgamation. Moreover, a direct contact between quicksilver and silver is a necessary condition for their amalgamation. A momentary contact in a muddy pulp is not always successful. The chloride of silver, however, goes into solution and unites easily with the quicksilver. Hence, in most instances, it is necessary to adopt a chloridizing roasting.

What a Chloride is, and how Chlorination is Effected.

9. The term *chloride* is applied to all compounds of chlorine with a metal or other radical. Chlorine is a greenish-yellow gas, an elementary substance, of 2.45 specific gravity, and of a peculiar and disagreeable odor. It is not found free in nature, but always in combination, principally with sodium, forming common salt. Metallic chlorides are of frequent occurrence. Chlorine is, for instance, combined with silver as horn silver, with copper as Atacamite, with lead as Kerasine, Mendipite, etc; also with quicksilver as Calomel.

10. To chloridize ore,—that is, to convert the metals into chlorides,—it is necessary to produce chlorine and to bring it in intimate contact with the ore particles. The cheapest material evolving chlorine is salt (chloride of sodium), and the only practical way of separating the chlorine from sodium is by substituting for it another substance for which the sodium has a stronger affinity. The cheapest ingredient for this purpose is sulphuric acid. The sodium being oxidized to soda, unites with the sulphuric acid, forming sulphate of soda, while chlorine is set free.

For the treatment of ores there are two principal methods of chloridizing. One is roasting the ore with salt in a furnace; the other is the " cold chlorination." Roasting, at first, when in the presence of salt, has an oxidizing effect, as there is then no sulphuric acid present to decompose the salt, and the heat alone would, if increased, volatilize and not decompose this. The sulphurets in the ore, under the influence of heat, lose a part of their sulphur as sulphurous acid gas; the other part of the sulphur oxidizes to sulphuric acid. As soon as this is formed it attacks the salt, and the chlorine, being set free, then acts on metals, metal oxides, sulphurets, arseniurets and antimonial combinations, forming partly metal chlorides and partly chlorides of sulphur, arsenic and antimony.

11. The other mode of chloridizing consists in the employment of cold chlorine gas with roasted ores, principally desulphurized gold ores, but of late, also silver ores. The chlorine must be produced here separately, and conducted into the cold ore by leaden or india rubber pipes. The ingredients are: salt, sulphuric acid and peroxide of manganese. Salt is first attacked by the sulphuric acid, and hydrochloric acid and sulphate of soda are formed. The hydrogen of the hydrochloric acid then combines with the oxygen of the manganese, and the chlorine escapes. A part of the chlorine unites with the manganese, but is decomposed again by sulphuric acid, so that all chlorine is expelled from the salt, leaving sulphates of soda and manganese in the gas generator. The chlorination of gold, unlike that of silver, is difficult to effect in a furnace (§ 38), for the reason that, if formed, the gold chloride is reduced back to the metallic state at a low, almost dark red heat. The difference between hot and cold chlorination is principally found in the fact that, while in the first way a great many base metal chlorides are formed, the cold chlorine combines principally with the free metal, with silver and gold; while the other metals, being oxidized, are not decomposed by the chlorine. Silver oxide, if present, is decomposed and chloridized.

Chlorination is also effected by chemical decomposition in the wet way, as practiced in the Mexican patio amalgamation, by mixing with the ore sulphate of copper and salt.

Means of Separating the Metal from Chlorine.

12. The chloride of silver can be melted without being altered; chlorides of gold and of plati-

num lose all their chlorine on being heated.
Chloride of iron exposed to air and heat, as is the
case in a chloridizing roasting, loses its chlorine
and is changed to iron oxide. The chloride 'of
copper gives up only a part of its chlorine. Heating alone has therefore no practical value for the
disengagement of chlorine.

The most effective way of separating the chlorine
from the metal is the application of another metal
for which the chlorine has more affinity. On this
property of chlorine is based the amalgamation of
silver ores, after a chloridizing roasting, in pans,
tubs and barrels, and the patio amalgamation.
The chloride of silver in the ore is decomposed,
and the silver set free during amalgamation in iron
pans by the metallic iron of the pan, or if quicksilver is charged at the same time with the ore, by
both the quicksilver and the iron. In the barrel
amalgamation the silver is disengaged by metallic
iron, and in the patio amalgamation by quicksilver.
In all these instances the silver, being deprived of
its chlorine, alloys with the quicksilver and forms
the amalgam.

On the same principle the metal is extracted
from soluble chlorides. The proto-chlorides are
all more or less soluble in water, except that of silver, which is quite insoluble. The chloride of
copper, in solution, is brought together with metallic iron, or conveyed over it. The chlorine of
the copper unites with the iron, and the copper
falls in a metallic state, ready to be melted into

bars, after being washed, pressed and dried. Indirectly, the silver is extracted from its chloridized state by dissolving the chloride in the hyposulphites of soda, of potash, or of lime. In these salts the silver chloride dissolves very readily, giving a clear solution of a very sweet taste, out of which the silver is precipitated by the corresponding alkaline sulphides as sulphide of silver.

The chloride of gold, obtained from the chlorination of gold-bearing sulphurets, is precipitated by sulphate of iron in such a way that metallic gold results, while the chlorine combines with a part of the iron.

13. The silver is easily obtained from the chloride by melting it with alkalies; for instance, with soda, potash or lime. The chlorine unites with sodium, calcium, etc., and the silver separates on the bottom of the crucible. If there is not a sufficient amount of the alkalies present, some silver will be lost. In most instances it is preferable to mix the artificial chloride with water and some sulphuric acid and granulated zinc, or zinc sheet if smaller quantities are being operated on. The chloride of silver by degrees changes its white color to a dark gray, being converted into the metallic state in a short time. It is reduced to metal by the nascent hydrogen. After the sulphate of zinc, which is formed and dissolved, has been washed away, the silver is pressed, dried, and, with addition of some soda and borax, melted into a bar.

In the same way as from a sulphate, silver can be precipitated by copper, after the chloride of silver has been dissolved in a hot solution of salt, as is done in Augustin's process. This is not practicable with the argentiferous solution of hyposulphite of soda.

By using sodium amalgam and iron filings, the silver chloride is instantly decomposed and silver amalgam formed.

The chloride of gold is precipitated in a metallic condition; also by the chloride of iron (Fe Cl), the consideration of which is important in treating sulphurets by chlorination.

II. ROASTING OF ORES.

14. The object of roasting is either to effect chemical changes, as required for amalgamation, smelting, etc., or sometimes also to reduce the hardness of the ore, in order to make it easier to crush. Roasting for the latter purpose, exposing the ore to the fire in large pieces, is more properly termed " burning." The beginning of smelting is under all circumstances beyond the limits of roasting; therefore all roasting furnaces in which the regulation of heat is so far out of the control of the roaster that a partial smelting would arise, are unfit for roasting. This is often the case with vertical furnaces. But although a partial smelting or clotting is not within the province of roasting, and in all instances is very injurious to the result of subsequent amalgamation or precipitation, it is nevertheless applied with much success on concentrated ore intended for smelting. By this process the loose sand assumes a compact form, the gases and wind penetrate the charge more easily, and the loss in metal is diminished.

If there is no necessity for effecting a perfect chemical change in the ore, or if roasting is required for smelting purposes, and a powdered form is not admissible, the ore is taken in larger or smaller pieces,—generally not below the size of a hen's egg,—and subjected to roasting either in open heaps, in kilns or in vertical or reverberatory furnaces. In roasting in heaps, the wood is first placed on the ground, sometimes surrounded by a wall two or three feet high, then the ore is put over it. Less frequently ore and wood are laid in strata. If there is sufficient sulphur in the ore, the burning will continue without addition of fuel for many days or weeks. It is evident that the result of such roasting is very unequal, the outside being more oxidized than the inside, the heat greater near the fuel than further off, etc. For this reason such ore is often roasted over several times.

In vertical furnaces, the ore is laid in strata alternating with fuel, or there are several fire-places outside the furnace so arranged that the flame is conducted by the draft into the furnace. A modification in construction and principle is the Hagan roasting furnace, in which the decomposition of superheated steam is a source of creating heat and a decomposing agent at the same time. The roasting is performed in a short time, and with proper ore and pieces of the right size the result is very satisfactory. It is also a cheap process, and is applied for roasting gold-quartz holding sulphurets, the amalgamation of which, without roasting, is defective. This kind of roasting would be also applicable as preparatory for amalgamating silver ores or for the chlorination process (§ 74).

In most instances with silver ores, a most perfect chemical change is a condition on which the result of extracting the silver depends ; and for this purpose the ore must be pulverized, in order to effect a perfect contact between ore particles, gases, and other substances which are mixed with the ore for certain purposes. The roasting of the pulverized ore is executed mostly in reverberatory furnaces; sometimes, also, in a kind of retort furnace, if the roasting should be done without the admission of air. There are also other furnaces lately introduced or tried, the description of which will be found hereafter.

In accordance with the intended mode of extraction, the ore is either roasted with an addition of charcoal powder, whereby the silver is reduced

to a metallic state,—a procedure of little practical
use,—or the ore is subjected to an oxidizing roasting, with the principal object of driving out
arsenic, antimony or sulphur, converting at the
same time the silver into a sulphate (Ziervogel's
process); or a chloridizing roasting is effected, that
is, roasting with salt.

A. Chloridizing Roasting.

15. In order to chloridize the ore, an addition
of common salt is indispensable. The salt furnishes
chlorine for that purpose, and is decomposed by
sulphuric acid. The sulphuric acid is created by
the decomposition of sulphurets present in the
ore (§ 10). . It follows that if silver ore is to be
roasted successfully with salt, there must be a certain percentage of sulphurets in it; otherwise no
sulphuric acid can be obtained, and consequently
no chlorination, or at least only a very imperfect
one, can be effected (§ 18).

Before introducing the ore into the furnace the
latter must be gradually heated up, which may take
ten to fifteen hours. When nearly red hot, a charge
of dry ore, mixed with salt, is brought on the hearth
through the roof and spread out equally by means
of a hoe. The fire is kept up moderately, but sufficient flame must be seen over the ore. The draft
is lessened by the damper, and the ore stirred diligently, but not continually. The intervals, however, must be short. In case the ore contains lead

and antimony, it is advisable to stir continually for at least three hours. The ore by degrees becomes red hot, and the burning of the sulphur is quite lively. One part of the sulphur, by the action of oxygen, is converted into sulphuric acid and combines with the metals, deprived of their sulphur or arsenic, to a sulphate. The period of the formation of sulphates is very important and requires some time before it is finished. If there is a large amount of sulphurets in the ore, the burning of the sulphur creates so much heat that the feeding of the fire must be stopped almost entirely for an hour or two, but must be resumed again as soon as it is perceived that the ore commences to cool. The workman stirs the ore, with a hoe or an iron rake, back and forward across the hearth, moving it from the bridge toward the flue and back. The formation of sulphates still continues with disengagement of sulphurous gas. The ore at the bridge is more exposed to heat than that on the opposite side, and the roaster is obliged to change the ore by raking it together into a long heap extending from the bridge toward the flue— not in the middle, but nearer the working door. By means of a shovel, six inches by twelve, on a long (12-foot) iron handle, the roaster takes the ore from near the bridge and transfers it toward the flue, putting it behind the ridge of ore until he reaches the middle of the furnace. He then takes the other end of the ridge and moves it toward the bridge. After this the stirring is continued in the usual way.

16. The sulphates react now on the salt, and decompose it under increased heat, setting the chlorine free. A mutual exchange takes place in part. Sulphate of lead changes into chloride of lead, which, volatilizing and coming in contact with air, loses one part of its chlorine and is reduced to a combination of oxy-chloride of lead. Sulphate of iron and sulphate of copper change also into chlorides. The copper chloride becomes volatile, colors the flame blue, emits chlorine gas and forms subchloride of copper. The chlorine, set free, decomposes the sulphurets and sulphates of silver, and creates chloride of silver. If, during the operation, lumps are formed, in case the ore was not dry enough or too much heat was applied in the beginning, they must be crushed to powder by a hammer-like iron instrument with a long handle. As soon as the chlorination begins, after three or four hours, a different smell, that of chlorine, will be observed. White fumes arise, and gases and vapors are evolved, consisting of sulphurous acid, chlorine, hydrochloric acid gas, chloride of sulphur, of iron and of copper.

The ore increases now in volume and assumes a wooly condition. Another hour's roasting will now finish the chlorination. This last hour's stirring requires a light red heat in order to destroy as much as possible of the base metal chlorides. If there is a great percentage of copper and other base metals in the ore, the roasting may require more time, in order to decompose the chlorides and sulphates, the presence of which consumes too much iron, and during amalgamation in barrels, increases the heat to such a degree as to cause an injurious division of the mercury into small particles and scum. The base metal chlorides are reduced by the iron and also amalgamated.

The changing of the cooler portion near the flue with the hotter part at the bridge must be repeated two or three times during the roasting process. When finished, after five or six hours, the ore is drawn out and discharged through the discharge-hole in the bottom. White fumes and gases are still arising.

The hoe, Fig. 1, is made of $\frac{1}{2}$-inch wrought iron, six inches high and eight inches wide. The rod or handle must be fifteen feet long at least. This would render the instrument heavy and tiresome to handle; it is therefore preferable to use a piece of gas-pipe, welding it together with the rod as represented in Fig. 1. The rake is generally of cast iron, and is shown in Fig. 2.

FIG. 1.   FIG. 2.

2

Necessary Amount of Sulphurets.

17. In times when the barrel amalgamation
was yet practiced in Freiberg (Saxony), long experience showed that a large amount of iron sulphurets was necessary in order to decompose the
amount of salt required for the chlorination. One
hundred parts of the ore were mixed with 150 parts
of borax glass, 100 parts of common glass and one
part of resin. This mixture, melted in an assay
crucible, gave a button of matt (sulphide of iron),
the weight of which was from 25 to 30 per cent. of
the original weight of the ore. If less matt was
obtained, the ore was considered too poor in sulphurets and more pyrites had to be added.

There is not much silver ore found in the State
of Nevada which would give 25 per cent. of matt
on the average; and as there is no pyrites to be
obtained for this purpose, the ore must be roasted
as it is. When starting the first amalgamation
works in Nevada, I found from six to eight per
cent. of sulphurets (different kinds) in the Com-
stock ore, which, after roasting, contained 88 per
cent. of its silver converted into a chloride. The
ore from the Rising Star mine (Idaho) had not over
8 or 10 per cent. of sulphurets, still there was 91
per cent. of chloride of silver found after roasting.
It is, however, very probable that from silver ores
containing a great deal of calc spar or heavy spar,
a less satisfactory result might be obtained by

chloridizing roasting, if no more than 6 per cent. of sulphurets should occur in them. Some copper-holding ores, especially if other base metals are present, and no sulphur (or very little), will give sometimes a good chloridizing roasting without any addition of green vitriol or other sulphur combination.

18. In treating ores entirely free from, or with a very small percentage of sulphurets, the want of sulphuric acid must be remedied by adding another substance. A cheap material of this kind is found in the green vitriol or copperas (sulphate of iron), of which 1½ to 3 per cent. is added when 8 to 10 per cent. of salt is used. The copperas is first calcined, in order to drive out its water of crystallization, by a gentle heat, and from the calcined article, not the crystallized, is taken the above percentage. This sulphate acts then on the salt the same as if it were created in roasting. The copperas is also added to arsenical ores free from sulphurets. But the percentage of green vitriol to be added depends also on the nature of the gangue. If there is a great deal of lime in the ore it takes up sulphuric acid, forming sulphate of lime, remaining in this condition through the process of roasting without being decomposed further. For this reason calcareous ore requires as much more green vitriol or iron pyrites as is necessary to transform all lime into a sulphate. Silica or quartz, if abundant, in the presence of steam, de-

composes some of the salt when red hot, forming
silicate of soda and hydrochloric acid, the importance of which is shown by the fact that gaseous
hydrochloric acid, in contact with metallic silver,
unites with it to a chloride. It behaves in a like
manner with sulphurets and arsenides, of which the
most are decomposed, forming chlorides, while
sulphur and arsenic escape combined with hydrogen.

Amount of Salt to be Used, and When.

19.   Ores containing from 80 to 100 ounces of
silver per ton should be mixed with 10 per cent. of
salt.   This is about the quantity considered necessary in the amalgamation works of Europe. Rich
ore is often roasted with 20 per cent. of salt. If
all the chlorine of the salt could be transferred to
the silver, an insignificant amount of salt only
would be required for ores containing 100 ounces
of silver—not more than $3\frac{1}{2}$ pounds to the ton; but
in consequence of the different ways in which
the chlorine decomposes and unites with base
metals and gases, the escape of chlorine from the
surface of the ore without coming in contact with
the silver, etc., a great deal more of the salt must
be applied.

The usual amount of salt used in the United
States for ores of the above value, is from 120 to
140 pounds per ton of ore; that is, from 6 to 7 per
cent. It is not advisable to take less than 6 per

cent. (§ 37), even if the ore be poorer. There are instances, however, where 91 per cent. of silver has been obtained by amalgamation from ores which were roasted with only 5 per cent. of salt. There was no natural chloride of silver in the ore when treated with 5 per cent. (Rising Star ore).

As the salt is not at all decomposed before the formation of sulphates commences, or only to a very small extent, it is also in this respect immaterial whether the salt is charged at once with the ore, or whether it is introduced two hours later, unless the ore is of such a nature as would bake easily on a little increase of heat. In other cases, however, it is obvious that, taking only 6 per cent. of salt, and employing only one man at a furnace, a perfect mixing in a short time, as ought to be done if the salt is charged after the sulphur is burned off, cannot be expected, and consequently a defective result will follow. It is therefore under such circumstances important to have the salt and ore introduced at the same time. The most perfect application of the salt is undoubtedly when both ore and salt are crushed together in the battery.

But a point of great importance is the time when the salt should be added, if other objects are in view. If the salt is added together with the ore, or after the sulphur is expelled and sulphates are formed, in every instance the base metals will take up their share of the chlorine, and therefore more salt will be required. But as the most of the chlorides are volatile, the salt is the means of getting rid of a great deal of the metals during the roasting, which in some instances is not very desirable. For instance, if a great deal of antimony and copper is in the ore, more or less chloride of silver will escape; sometimes, however, only a small percentage. Treating the ore with salt from the beginning, or adding it two hours after the beginning, the result is the same.

A different result is obtained if the salt is added after all the base metals are desulphurized and oxidized. Some base metals, as antimony and arsenic, will be volatilized and thus gotten rid of, but not in so large a proportion as if chloridized. Iron and copper remain entirely in the ore, while both are volatile as chlorides. The roasting must be continued at a light red heat till all sulphates are decomposed and the metals oxidized. Applying the salt after the dead-roasting, the effect differs from the above so far, that the base metal oxides now are not chloridized, or only to a small extent, while the silver alone (some of which appears to be changed to a metallic state, the most, however, remaining as a sulphate) will be chloridized. But in order to effect this chlorination, from 1 to 2 per cent. of green vitriol must be added in order to accomplish the decomposition of all the salt. The copper is lost with the tailings unless smelted, or extracted by diluted sulphuric acid.

Permanent Stirring not Essential.

20. In roasting the ore with salt, a continual stirring to the end of the process is not a necessary condition for obtaining a good result. This depends partly on the time and partly on the nature of the ore. As long as the ore is not uniformly heated, a diligent stirring is important. The ore in the corners is too often neglected while the sulphur is burning, and the exposure of a fresh surface to the oxygen of the air requires also constant work; but as soon as the smell of the chlorine is perceptible, the stirring can be carried on at intervals of from eight to ten minutes. The chlorine which is evolved in the mass (§ 23) has better opportunity to act on the metals than if constantly stirred, whereby more chlorine escapes up the chimney without producing any effect. This was proved by a comparison of the work of two furnaces. A revolving furnace had a speed great enough to let the ore drop constantly through the flame and air, while the common furnace was managed by only one man, and stirred at intervals. Mr. Atwood found 15 per cent. less chloride of silver in the roasted ore from the revolving furnace. The blame is not with the revolving furnace, but with the speed. It proves, however, that, being constantly exposed to the air, the chlorine escapes with less effect than in the common furnace, where the ore is allowed to rest for ten or fifteen minutes, and the evolved chlorine, being in contact with the particles while passing through the mass, is permitted to form combinations. O'Hara's mechanical furnace, in which the ore is comparatively but little stirred, gave 91 to 94 per cent. of chloride of silver. A mixture of ore, sawdust and salt, formed into bricks and calcined, showed the silver as a chloride through the whole mass, where, as a matter of course, the inside did not come into direct contact with air. Constant shoveling is necessary with ore of such a nature, as it would bake if not stirred.

Signs of a good Chloridizing Roasting.

21. A good chloridizing roasting should give over 90 per cent. of the silver converted into chloride of silver, and show as little as possible of base metal chlorides. To ascertain the amount of chloride of silver at the end of the roasting, it is necessary to make two assays. About one ounce and a half is taken out of the furnace and allowed to cool. Two one-half ounce assays are weighed out, and one (No. 1) prepared for the fire assay as usual. The other half ounce (No. 2) is introduced carefully into a small filter in a glass funnel. The filtering paper must project about one inch above the ore. A solution of hyposulphite of soda is then poured over the ore in the filter, and this is continued as long as a precipitate is obtained on adding a solution of sulphide of sodium to the filtored liquid. This is best tried in a clean glass tube. If, after filtering or leaching, the addition of sulphide of sodium to the leach does not produce a precipitate, or only a very slight one, so that the liquid assumes only a little darker color without losing its perfect transparency, the assay is leached with warm water and the filter taken off, put into a porcelain dish or like vessel, and dried, by applying heat with an alcohol lamp. The filter can be removed and burned above the sample. When dry, the ashes of the filter and the sample are fluxed like the other half ounce, and both crucibles placed in the assay furnace.

It is supposed that the examination of the roasted ore is undertaken when it is thought that the chlorination is nearly finished; otherwise too much sulphate of silver would be taken for a chloride. Should it be required to ascertain the amount of pure chloride of silver formed, at any time during the roasting, three assays, of half an ounce each, should be made. For this purpose half an ounce is weighed out for the usual fire assay. Another half ounce must be leached with hot water, by which the sulphate of silver is dissolved, and the third sample is treated with hyposulphite of soda, as described above. Comparing the results of the three assays, it is easily found how much of the original amount of silver was turned into a sulphate, and how much into a chloride.

Hyposulphite of soda is a crystallized salt, of which five ounces may be dissolved in a quart of cold or warm water.

For sulphide of sodium, take five or six ounces of soda or soda-ash, melt in a crucible, and when liquid introduce two ounces of sulphur (brimstone), at intervals, in small pieces, giving time for the boiling up to subside. Pour out on an iron plate and dissolve in water. It takes several hours before the solution appears perfectly clear, showing a yellow color above the black sediment. When drawn off it is ready for use.

The operation takes less time if the hyposulphite solution is used in a hot state. All chloride of silver, and also sulphate of silver, if present, is

2*

dissolved by the hyposulphite and carried off, beside the base metal chlorides. The two assays, when ready, are compared, and the difference shows the silver which was converted into a chloride. For instance, if No. 1 assayed 83 ounces per ton, and No. 2 from the filter 4 ounces, the difference, 79, is that part which became chloridized. That is,

$$83 : 79 = 100 : x = 95 \text{ per cent.}$$

22. Toward the end of the roasting very little, if any, sulphate of silver will be found in the ore; but even if a small percentage of it should remain, it may, for the purpose of amalgamation or extraction, be considered equal to chloride of silver; for as soon as it dissolves in water, it becomes a chloride, precipitated by the salt, of which a part is always yet found undecomposed in the ore. To obtain a general idea of the amount of soluble base metal chlorides and sulphates, it is sufficient to put a small sample of about half an ounce on the filter as before, and to leach it with hot water. The leach obtained is tried again with the sulphide of sodium. A thick precipitate shows that a large amount of soluble chlorides is in the roasted ore. If a reaction of copper is expressly desired, ammonia should be used in place of the sulphide of sodium. In presence of much iron the precipitate will appear brown. This precipitate must smell strongly of ammonia. If copper is present, a clear blue liquid will be seen above the iron precipitate after some time; or the whole may be brought on a filter to separate the liquid from the precipitate.

Means of Destroying Base Metal Chlorides.

23. It is very difficult to get rid of all the base chlorides. They are formed under the action of chlorine and hydrochloric acid. The most of the metal chlorides are volatile, and a part is carried off through the chimney. Another part of the chlorides gives off some of its chlorine, whereby sulphates, undecomposed sulphurets, antimonates, arsenates and free oxides, are chloridized. Chlorides which are disposed to transfer chlorine to other metals in combination with sulphur or arsenic, are: the proto-chloride of iron and of copper, the chlorides of zinc, lead and cobalt. When in this way the most of the metals are chloridized, the base metals, principally iron and copper, are losing their chlorine gradually, being first converted into sub-chlorides and then into oxides. The roasting for this purpose must continue with increased heat, even when the chlorination of the silver is finished. At an increased heat, the base metal chlorides lose their chlorine, while the chloride of silver remains undecomposed, unless a very high temperature should be applied. This process requires a long time, consequently also more fuel. The decomposition of these chlorides is greatly assisted by the use of 5 to 6 per cent. of carbonate of lime in a pulverized condition. Lime does not attack the chloride of silver, but it is not advisable to take too much of it, as it would interfere to some degree with the amalgamation. The pulverized lime rock must be charged toward the end of the roasting. First, two per cent. is introduced by means of a scoop, the whole well mixed, and then examined either with sulphide of sodium (§ 22) or in the following way:

A small portion of the roasted ore is taken in a porcelain cup or glass, and mixed with some water by means of a piece of iron with a clean metallic surface. If the iron appears coated red with copper, some more lime must be added. In place of iron,—especially if no copper, but some other base metal is present,—some quicksilver is mixed with the sample. In the presence of base metal chlorides, the quicksilver is coated immediately with a black skin.

When endeavoring to expel the base metals by heat, the loss of silver, in presence of much antimony, lead and copper, should be investigated very carefully. Under certain circumstances it is not uncommon to find a loss of even 50 per cent. of the silver, if the chloridizing roasting is carried on at a high heat for a great length of time. The loss increases with the duration of roasting and with the degree of temperature. When such ore is under treatment, it is necessary to take samples during the roasting, and to examine the same for the amount of chloride of silver, and also for its loss, and to stop roasting when the highest percentage of chloride of silver is obtained, without reference to the condition of base metals.

Steam Decomposes Base Metal Chlorides Effectively.

The formation of base metal chlorides can be avoided by a proper but more expensive roasting (§ 33). It requires, first, an oxidizing roasting, with the application of steam. This roasting must continue until all the metals are desulphurized and converted into oxides. When this is accomplished, salt and green vitriol are added, and the roasting continued until all the silver is chloridized.

There is also a very good way of getting out a great deal of the base chlorides of the ore before the silver is amalgamated or extracted, by leaching the ore with hot water (§ 77).

Application of Steam in Roasting.

24. The application of steam in roasting is advantageous, for the reason that hydrochloric acid is created by the decomposition of chlorides; which in turn decomposes the sulphurets. The hydrogen decomposes also the chloride of silver, which, upon being reduced to metallic condition by its affinity for chlorine, in turn decomposes the hydrochloric acid. The silver may thus change repeatedly from a metallic condition to a chloride, while the base

metal chlorides are reduced to oxides, and in that
state do not interfere with the amalgamation or
precipitation. The application of steam, however,
requires a great deal more fuel during the roasting. Taking the moisture of the fuel into consideration, there is no roasting done without steam,
although with a limited quantity.   .

Silver Ore, Containing Lead, Unfit for a Chloridizing Roasting.

25. Lead has a bad influence in amalgamation
and precipitation, and even in the roasting itself,
causing a baking of the ore at the slightest undue
rising of the temperature. The chloride of lead
amalgamates easily, especially in iron pans. Ores
with 8 to 15 per cent. of lead still allow of a successful roasting. A part of the formed chloride of lead
escapes in gaseous form, another part is reduced
by degrees to oxy-chloride of lead. This latter
combination goes mostly into the amalgam. If
there is more lead in the ore than 15 per cent., it
gives sometimes, according to its nature, as much
as 85 per cent. of silver, and the retorted amalgam
is submitted to cupellation in order to separate the
lead. In Hungary (Offenbanya), black copper, containing, besides the silver, 10 per cent. of lead,
is subjected to a chloridizing roasting. The
pulverized copper is mixed with 12 per cent. of
salt, 1 per cent. of green vitriol, and 3 per cent. of
saltpetre. The saltpetre oxidizes the lead to a sul-

phate, which is not affected in the subsequent amalgamation in barrels.

Difference in Roasting Ore for Pan Amalgamation, as compared with that for other Modes of Extraction.

26. The roasting of silver ores, if imperfect, will give a better result by amalgamating in an iron pan, than in wooden barrels or by precipitation. This is due to the better decomposition of sulphates and undecomposed sulphurets under the grinding muller. The roasting for pan amalgamation is therefore less delicate. However, when once at work, it is always better to do the roasting properly; but it is not necessary to sift the ore after roasting in order to separate the lumps from the mass, as is done with the barrel amalgamation, except to prevent nails from coming into the pan. The formation of such lumps, however, must be avoided as much as possible. Imperfect roasting in the presence of base metals, gives in pans always a low fineness of bullion.

Examples of Local and Different Roastings.

27. *Roasting of Silver Ores in Freiberg (Saxony).* The amalgamation process, and consequently this kind of roasting, was given up long ago at Freiberg; but the method of roasting as performed

there is nevertheless very interesting. The ore
subjected to roasting consisted of silver glance,
brittle silver ore, ruby silver, metallic silver,
fahl ore, bournonite, zincblende, sulphide of antimony, iron, copper, and nickel pyrites, and of
gangue, viz: quartz, calc, brown, heavy, and fluor
spar. It contained from sixty to one hundred
ounces of silver per ton.

The dry crushed ore was first spread on a platform ; on this a layer of damp ore, from the wet
concentration, was laid, and then 10 per cent. of
salt. This order was repeated from six to eight times.
The stratified mass was mixed thoroughly by means
of shovels and a coarse sieve. This mixture contained from 9 to 10 per cent. of moisture. For
this reason, after a charge of 450 to 500 pounds
was introduced through a hole in the roof, the fire
was kept very low, in order to dry it at a dark red
heat, and the ore was diligently raked by two men,
working alternately. As soon as the decrepitation
of the salt ceased, the ore was ridged from the
bridge toward the flue, through the middle of the
furnace, and the formed lumps broken up by iron
hammers attached to long handles. After this was
done, the heat was increased, whereby the ore,
under constant stirring, assumed a red hot condition, and the sulphur commenced to burn quite
lively. This stage was reached in two hours from
the beginning. The desulphurization commences
with the burning of the sulphur, creating a temperature sufficiently high to continue the roasting

without fuel for some time. Fumes are evolved, consisting of steam, antimony, sulphurous acid, arsenic, etc. This desulphurization takes again two hours, the workmen all the time raking and changing the hotter part of the ore at the bridge with the cooler at the flue. The temperature is now raised to a light red heat, the ore increases in volume, emitting chlorides of metals, chlorine, hydrochloric acid, etc. The formation of the chlorides progresses rapidly, and is finished in three-quarters of an hour. The charge is then drawn out. A too long roasting would not give an equally good result, as some silver might be decomposed to the metallic state, which is not so readily amalgamated as the chloride.

In what Condition the Metals are after Roasting.

28. After roasting, the silver is found almost entirely converted into a chloride, but a small part may remain as a sulphate ; antimonate of silver also is formed. The iron is converted into an oxide, some into sulphate and arsenate and basic chloride. The copper appears also oxidized, with some sub-chloride and less chloride. Lead remains principally as a sulphate and basic chloride. Zinc is oxidized. Antimony is found as antimonates combined with oxide of antimony, and with other basic oxides. Nickel and cobalt remain as oxides, chlorides and arsenates, and the arsenic is found as an

arsenate combined with other metals. Besides
these metal combinations, there is undecomposed
salt, sulphates of soda, of lime, etc. The charge
loses about 10 per cent. of its weight, of which a
part is regained from the dust chambers.

Antimonate of silver is considered the cause of
the loss of silver in roasting, as it is not decomposed in barrels, and consequently not amalgamated. There have been no experiments made as
yet to determine whether antimonate of silver is
decomposed in iron pans during the amalgamation
or not. Probably it is not, as the natural combination of antimony, lead and silver, cannot be
amalgamated in pans.

29. *Roasting of Argentiferous Copper Ores.* At
Arivaca (Arizona) the silver ores from the Heinzel-
mann mine consisted principally of silver copper
glance (stromeyerite) with 51 per cent. of silver;
fahl ore with from 2 to 15 per cent. silver, containing also some quicksilver ; zincblende, galena and
other decomposed argentiferous copper ores. On
an average the ore contained from $150 to $200 per
ton, and from 10 to 15 per cent. of copper. After
crushing, the ore was spread on a platform covered
with 8 per cent. of salt, and mixed thoroughly by
means of shovels. Eight hundred pounds of it, as
a charge, were introduced through the roof of the
furnace (which was constructed entirely of adobe),
spread on the hearth, and, at a dark red heat,
stirred for two hours, at the end of which time the

flame was colored an intense greenish-blue, and considerable fumes were emitted. The raking continued for an hour and a half more at an increased heat, and during this time the ore was moved three times from the bridge to the flue and back. A sample taken from the furnace at this time, put on a canvas filter, wet with salt solution and leached with a hot concentrated solution of salt, gave a clear liquid, which, diluted with water, showed a strong white precipitate of chloride of silver, mixed with antimony and lead; but the quicksilver, treated with the same sample of ore and water, was cut and blackened to a high degree. For this reason, from 5 to 6 per cent. of pulverized lime was thrown into the furnace by means of a scoop, as much as possible over the whole surface of the ore, and then raked and stirred diligently in order to finish the mixing in the shortest time. After four hours from the beginning, the temperature was raised to a light red heat for half an hour, and the roasting was finished. There were yet a great many base metal chlorides in the ore, but as metallic copper was used in the barrels, the silver turned out always over 900 fine. The loss of silver was 12.5 to 13 per cent.

30. *Roasting of Copper Matt.* In smelting argentiferous copper ores, the process is sometimes regulated to produce a sulphide of copper, containing silver and base metals, as antimony, arsenic, zinc, iron, etc. This sulphide of copper, or copper

matt, was roasted formerly and smelted again to
produce black copper; that is, impure metallic
copper. For the purpose of extracting the silver
therefrom, the copper was melted together with a
certain percentage of lead, and the latter, with the
silver, extracted by liquation and cupelled. The
remaining copper contained still some silver and
lead, and the process was a very lengthy one before finished. To avoid the liquation, the copper
matt was treated by amalgamation, and the silver
extracted at once. For this purpose the matt was
crushed and sifted, and the coarse part ground.

Of this powdered matt, 300 pounds are charged
in a double furnace, of which the upper hearth
prepares the ore by a moderate roasting, while the
lower one finishes the operation at a higher temperature. In each of the hearth-departments the
matt is treated for two hours and a half. Silver
and the other metals are first converted into sulphates and then mostly decomposed to oxides, but
the silver remains for the greatest part in metallic
condition. The matt is drawn out, mixed with 8
per cent of salt and 12 per cent. of lime, and with
salt water into a paste, which is allowed to rest for
twelve or fourteen hours. The paste is then dried,
powdered between rollers, and again roasted two
hours and a half. During this process, samples
are taken and mixed with water and a few drops of
mercury. If this appears coated bluish, it proves
the presence of metallic salts, and some more lime
must be added; if after this the quicksilver re-

mains perfectly white,—parting, however, in many minute globules,—it proves that too much of the lime was used, and in this case some of the first roasted matt is added.

The purpose of wetting the roasted ore, as above described, is the formation of chloride of silver. As there are always sulphates of the metals present after the first roasting, they decompose the salt, and the chlorine acts on the metallic silver. This process is not perfectly finished, and therefore the second roasting.

31. *Roasting of Black Copper.* The black copper obtained from smelting (in Schmoellnitz, Hungary) contains from 110 to 150 ounces of silver per ton, and 85 to 89 per cent. of copper. To pulverize this it must be made red hot in a reverberatory furnace and crushed while red hot. The powder must be sifted and then ground fine. The pulverized metal is then mixed with 7 to 9 per cent. of salt, and roasted in the usual way for six to six and a half hours, in charges of 400 pounds each. No green vitriol is added for the purpose of decomposing the salt ; and as there is not more than from $\frac{1}{2}$ to 1 per cent. of sulphur in the black copper, the salt decomposes through direct action on the copper. First, chloride and subchloride of copper are formed. The copper chloride transfers chlorine to the metallic silver, and is reduced to a sub-chloride.

In other places iron pyrites are added to the

black copper, by which the chlorination is pro-
moted. At Oriklowa (Banat) 5 per cent. of iron
pyrites and 12 per cent. salt are used, roasting
twelve hours. The loss of silver is 7 per cent., and
of copper 3 per cent., during the roasting. The
expense of roasting is $7.30 per ton.

32. *Roasting of Silver Ore at Flint, Idaho Territory, in O'Hara's Mechanical Furnace.* The ore
from the Rising Star mine, at Flint, contains argentiferous fahl ore, miargyrite, ruby silver, zincblende, galena, iron pyrites and sulphide of antimony. On an average the ore paid between $90
and $100 per ton, containing some gold. The
gangue is quartz. It is crushed through sieves with
forty holes to the inch, together with 5 per cent.
of salt. The furnace (§ 57, Fig. 8) is charged continually by machinery at one end, *a*. The ore is
moved by degrees forward, and arriving at the first
fire-place, *c*, commences to disengage sulphur.
Between this fire-place and the second, which is on
the opposite side, between *c* and *d*, the chlorination
begins at an increased heat. The flame shows
partly a blue color, originating from chloride of
copper, and white fumes are also evolved. Between the second and the third fire-place, *d*, the
chlorination is finished at a light red heat. From
the cooling hearth, *e*, the roasted ore is continually
discharged on the dump, *f*. It takes six hours before the ore from the feeding place, *a*, arrives at
the dump. Although not more than 5 per cent. of

salt is added, the roasted ore contains about 90 per cent. of the silver converted into a chloride. The gases, containing free chlorine and chloride combinations, emitting chlorine (§ 16), being in contact with the surface of the ore while passing over it for a space of eighty feet, have a chloridizing influence on it, replacing thus a certain amount of salt.

These three furnaces roasted twenty tons of ore in twenty-four hours. The expenses were as follows :

For wood, five cords, at $5................	$ 25 00
For four men at the furnaces, at $4........	16 00
For two men bringing in wood, etc., at $4..	8 00
For one man as watch in the night.........	4 00
For blacksmith work...	5 00
For 2,000 pounds of salt, at 8c............	160 00
Total expense..........................	$218 00

or $10.90 per ton. The capacity of the three furnaces is calculated for more than twenty tons. Each one could easily treat ten tons of the Rising Star ore in twenty-four hours. The roasted ore was treated by amalgamation in pans, applying the " leaching process."

Taking moderate prices of salt and labor into calculation, instead of the " Flint Tariff," the roasting expense in O'Hara's furnaces would not exceed $5.50 per ton, provided three furnaces were at work.

33. *Roasting of Silver Ores for the Patera Process.* The ores treated by Patera's process (§ 71) at Joachimsthal are remarkable for the numerous mineral species occurring in the ore. Among these may be mentioned silver, lead, different compounds of copper, bismuth, iron, uranium, nickel, cobalt, etc., sulphur, arsenic and antimony. Before the introduction of Patera's process, the extraction of silver, on account of so many base metals, was very difficult. The success of Patera was not so much in adopting hyposulphites, as proposed by Percy and Hauch, but in his modification of the roasting process, by which only the silver was converted into a chloride.

The pulverized ore is placed in the furnace, in charges of from four to five hundred pounds. First quite a moderate heat is applied, and gradually increased, but not so much as to induce clotting. As soon as the ore appears red hot, steam is admitted, with about four pounds pressure to the square inch. As the steam consumes heat, more fuel must be used to keep up a red heat. The ore must be constantly raked during the whole period of this roasting. It takes about four hours to finish this process, after which the ore is drawn and permitted to cool. The iron appears now as an oxide—also the copper ; some sulphate of copper is also present, and the silver is principally in the state of a sulphate.

The oxidized ore is now ground finer, mixed with from 5 to 12 per cent. of common salt, and, at the same time, with 2 to 3 per cent. of calcined green vitriol. This mixture is spread upon the hearth in the furnace, and subjected to a second, now chloridizing, roasting. It takes about an hour before a

red heat is reached. Steam is then introduced, as
above, under continuous stirring. The fire is gradually increased, and the roasting finished within from
five to eight hours, according to the value of the
ore. There are condensing chambers for catching
volatile metals and ore dust. They áre of importance if rich ore is treated, and without this con-
trivance several per cent. of silver would be lost.
When finished, the ore is drawn and allowed to lie
undisturbed for some time, after it has been moistened. In this condition the chlorination continues. The application of steam causes nearly
twice the consumption of fuel, but it has been shown
that, by the steam, hydrochloric acid is formed,
whereby arsenic and antimony are expelled, and at
the same time the chlorination of the silver greatly
favored. Over one ton of coal and one-half cord
of wood are consumed for each ton of ore roasted
in this way.

34. *Roasting for Augustin's Process.* To this
process principally matt is subjected. The process
requires a chloridizing roasting. The ore, if not
itself rich in sulphurets, is mixed with iron pyrites, slag, lime, etc., and smelted. The molten
sulphide of iron takes up the silver and deposits
itself below the slag on the bottom of the furnace.
The silver is thus separated from the earthy part
and concentrated in the matt. Argentiferous copper ores are likewise smelted for the purpose of
obtaining argentiferous copper matt. The matt is

3

then finely pulverized, and 400 pounds of it are introduced into the furnace. The roasting goes on now in the usual way, by starting with a moderate temperature, gradually increasing it, exposing every portion of the ore to the intense heat by frequent stirring, etc. At the end of eight hours the roasting is generally completed, the matt looks dark and earthy, and no fumes of sulphurous acid can be perceived. The roasted stuff is now drawn out and permitted to cool. After this the matt powder is ground finer, sifted or bolted, and given back to the furnace.

Four hundred pounds are mixed with 5 per cent. of salt. Sulphates are present, but oxides of metals predominate in the mixture. The formation of the chlorides commences immediately, and the roasting is concluded after one or two hours. The temperature in this second roasting is kept low, as the smelting of the chloride of silver must be prevented—for when melted the chloride of silver dissolves with more difficulty in a salt solution. In other places, after the first roasting, the mass is not taken out and re-ground, but, when finished, only a portion is drawn, mixed with from 1 to 6 per cent. of salt, according to the purity of the matt, charged again, mixed with the balance of ore in the furnace, and roasted for one-half to three-quarters of an hour.

35. *Roasting of Silver Ores for the Chlorination Process* (§ 74). It is not absolutely necessary for this process to have the ore roasted with salt, but it has been found that, on account of different earths, an addition of 1 or 2 per cent. of salt produces a better result. This process extracts copper, gold and silver, each of which is obtained separately; but it makes a difference in roasting, whether the copper is intended to be saved or not, as in many localities the copper is at present of no value, or the old iron for precipitation is too expensive.

The ore is crushed dry through a sieve of forty holes to the running inch. Some ore allows also thirty holes to the inch. In case the ore contains so much clay or talc that no leaching is admissible, the ore is crushed wet, separated from slime, and dried. Eight hundred to one thousand pounds are charged, and, according to the quality of the ore, the heat raised quickly or slowly to a bright red heat. This is an oxidizing roasting, consequently much stirring is required. Ores with but few sulphurets appear sufficiently well roasted after three hours; other ores containing an abundance of sulphurets, take from five to six hours and more before all sulphur, arsenic and antimony are expelled; and the arsenates and antimonates formed by the two last are not volatile. When desulphurized, 1 or 2 per cent. of salt is thrown in the furnace and mixed with the ore as intimately as possible. Three-quarters of an hour after the addition of the salt, and with only a moderate heat,

the roasting is finished.   Ores, not rich in sulphurets, may be mixed with salt when charged.

If it is intended to extract the copper also, this
must be transformed into chloride of copper.   To
accomplish this, two things must be observed,—
first, no oxide of copper should be formed during
the roasting ; and second, more salt must be used.
Stoichiometrically, each pound of copper requires
1.84 pounds of salt to form a chloride, provided
all chlorine is taken up by the copper ; but as
this is not the case, as a great deal of chlorine is
also absorbed by other metals, etc., it follows that
at least two pounds of salt, if not more, must be
taken for each pound of copper in the ore.   For
most localities such a quantity of salt could not be
used on account of the difficulty in obtaining it.
It may be mentioned here that if the brine, remaining after the copper has been precipitated by iron,
should be condensed by evaporation, exposing it to
the heat of the sun, which might be practicable on
the Pacific Coast in the dry season, the condensed
salt, consisting of chloride of iron, could be added
to the ore as a chloridizer, whereby a considerable
percentage of salt would be saved.

In roasting with salt with reference to extracting
the copper, the ore is first roasted for itself at a
low temperature, so as not to decompose the sulphates by a bright red heat, but long enough to
decompose all sulphurets..   When this is accomplished, the salt is introduced and the roasting
finished in one-half to three-quarters of an hour
thereafter.

36. *Roasting of Silver Ores in a Long Furnace at San Marcial, Sonora, Mexico.* For the purpose of roasting silver ores for the Solving and Precipitation Process, there are several long furnaces (thirty feet long) built up by Mr. O. Hofmann, in Sonora. Using long furnaces is found a great economy in every respect. A great advantage results also from moving the ore, at intervals of from one to three hours, from one hearth to the other. By this means it is impossible, even with careless roasters, to find raw ore in the finished charge, as would happen under such circumstances if the corners of single roasting furnaces were not very carefully attended to during the roasting.

The furnace at San Marcial is sixty feet long. The plan was made by Mr. Graff, who superintended the reduction works. It is a level hearth sixty feet long, representing six furnaces, each ten feet long, parted only by the projecting wall inside, as shown by v, Figs. 6 and 7 (§ 49). There are six working doors on one side, and, on account of the length, an auxiliary fire-place is placed at the back side, before the last two hearths. Each hearth contains 800 pounds of ore, and is attended by two Yaque Indians at a time, stirring alternately. These twelve Indians perform all the work about the furnace, carry the wood from the adjoining yard, split what is too thick, carry out the ashes, etc. The ore on the first hearth, nearest to the fire-place, has always a light red heat, which decreases with the distance from the bridge, so

that the fifth appears quite dark if not assisted
by the second fire-place. After stirring one hour
on all six hearths, the charge from the first is
drawn out through a door in the rear, on a large,
smooth platform, and immediately spread by means
of shovels in a layer one or one-and-a-half inches
thick, so as to have it cool enough for transporta-
tion to the sifting apparatus after the lapse of one
hour. As soon as the hearth is cleared, the ore on
the second hearth is moved over to the first, then
that on the third to the second, and so on, till
the sixth hearth remains empty and is charged
through the funnel in the roof with a new charge.
It will be seen that 800 pounds of ore are drawn
out every hour, and that each charge is exposed to
the fire for six hours. It is thus evident that,
being moved six times from one hearth to the other,
the ore arrives perfectly prepared to the finishing
heat. After roasting and sifting, the ore is amalgamated in pans, but as it contains some carbonate
and sulphuret of lead the amalgam is charged with
base metals, so much that refining by cupellation
is necessary. From 8 to 10 per cent. of salt is
added and mixed with the ore before it is charged.
Preparations are made now to introduce the solving and precipitation process, if successful on that
kind of ore.

The above arrangement, employing so many
hands, is considered by Mr. Graff a local necessity; the Indians are cheap, but not very attentive,
laborers.

According to an analysis made by Mr. Graff, the roasted ore from single furnaces (treating $100 ore) contained 5 per cent. less of chloride of silver than that from a long one. Long roasting furnaces are especially adapted for roasting sulphurets containing gold. Concentrated sulphurets, or ore containing an abundance of sulphurets, allow the use of a very long furnace, with only one fire-place, on account of the heat created by the burning sulphur.

The roasting expenses at San Marcial, with the furnace sixty feet long, are as follows :

```
24 men, day and night, at 50c...............$12 00
2 cords of wood at $3.......................  6 00
8 per cent. of salt=1,536 ℔s. at 2c......... 30 72
Repairs, etc................................  3 00
                                            -------
        Total expense on 9.8 tons.............$51 72
```

or $5 27 per ton.

A furnace thirty feet long, with the same kind of laborers, 800-pound charges, drawing every two hours,—that is, 4.8 tons in 24 hours, shows the following expenses:

```
8 roasters at 50c..........................$ 4 00
1½ cords of wood at $3.....................  4 50
8 per cent. of salt at 2½c................. 19 20
Other expenses.............................  2 00
                                           -------
        Total.............................. $29 70
```

or $6 18 per ton.

37. *Roasting in Stetefeldt's Furnace at Reno, Nev.* The mechanical part of the roasting itself, in this furnace, is the simplest of all, and also the shortest. The finely pulverized ore, mixed with salt, is sifted continuously by a mechanical arrangement into a shaft. This shaft, a, (Fig. 9, §58), is about twenty-five feet high, and heated by two fire-places provided with grates. The ore, falling through the heated shaft, undergoes chlorination,—a process requiring only a few seconds. After the roasted ore has accumulated on the bottom of the shaft to the amount of about 1,000 pounds, it may be drawn out. The amount of salt needed for chlorination, varies according to the ore; generally about 6 per cent, or 120 pounds to the ton, is taken ; or even less, especially in treating poor ores, when half of that amount may be sufficient in most cases. A furnace having a capacity of from fifteen to twenty tons in twenty-four hours, consumes from two to three cords of wood. In twenty-four hours there are employed: Two men attending the feeding and conveying machinery, three firemen, and three men to draw and cool the roasted ore. As the latter three have time enough to carry the ore to the pans, only half of their time should be charged to the roasting expenses. According to these figures, the total expense of roasting, in Reno, is not more than—

For labor of 6½ men at $3	$19 50
For wood, 2½ cords at $6	15 00
Salt, 1,800 pounds at 1½c	27 00
Total expense on 15 tons	$61 50

or $4 10 per ton. From 88 to 92 per cent. of the

silver contained in the ore is converted into a chloride. Of the dust in the dust-chambers, the silver was found in the state of a chloride up to 96 per cent. [See *Scientific Press*, 1869, p. 377].

It is evident that with an improper treatment of the fire, by using too much or too little fuel, a less favorable result would be obtained. In the first place, if the temperature is kept too high, a part of the chloride of silver is reduced to a metallic state, which, for the purpose of amalgamation, is not so very injurious (§ 8); but the metallic silver is a total loss with the solving process. On the other hand, if there is not sufficient heat, some of the sulphurets may remain undecomposed. In either case the responsibility is with those in charge of the furnace; but there is nothing easier than to keep up a proper and uniform heat in Stetefeldt's furnace, there being no other hand-work about it, and all the attention of the fireman being directed to this single point.

Stetefeldt's furnace has been compared sometimes with Gerstenhoefer's shelf furnace. In reference to this the *Engineering and Mining Journal* says :

" Since the discovery and exploration of the numberless mineral deposits in the Western States and Territories, no branch in metallurgy has received so much attention as the process of roasting ores of all descriptions. One can hardly look over a file of Mining Journals, or newspapers from some mining district, without finding descriptions

3*

of new devices for roasting ores,—all of .which
claim to surpass everything else in this line
which was known before. The devices are as
strange as they are many, and much time and
money has been wasted to test impracticable inventions. Indeed, the high expense which the
roasting in the old reverberatory furnace entails,
was a strong inducement to invent some cheaper,
and at the same time more effective, method.
This is especially of importance where silver ores
are found, which require a chloridizing roasting
preparatory to their amalgamation. In such cases
the expense of roasting is frequently more than
one-half of the total expense of reduction, and
consequently low-grade ores cannot be worked
with a profit. · But in spite of the necessity to
adopt some improved and more economical process of roasting, it has been extremely difficult to
introduce two inventions, which are based upon
the most simple and rational principles—so simple, indeed, that it seems impossible to simplify
them any more. We speak of the Gerstenhoefer,
or Terrace furnace, first introduced about six years
ago at Freiberg, and the Stetefeldt furnace, in-
vented three years ago at Austin, Nevada, but first
introduced for regular working at the mill of the
Nevada Silver Mining Company, near Reno, Nevada, in October of last year. The nature of these
inventions can be expressed as follows:

    Gerstenhoefer discovered that sulphurets are
completely roasted or oxidized if they fall against

a current of hot air rising in a shaft, which is filled with shelves, so as to check and retard the fall of the ore particles at certain intervals.

Stetefeldt discovered that silver ores, no matter in what combination the silver occurs, mixed with salt, are completely chloridized if they fall against a current of hot air rising in a shaft with no obstructions whatever to check or retard the fall of the ore particles.

It is a matter of course that in both cases a certain degree of fineness is required for the ore to be treated, and that a much coarser material can be successfully roasted in the Gerstenhoefer furnace than in Stetefeldt's.

We do not intend to enter here into a detailed description of the Gerstenhoefer furnace, since that invention has been frequently laid before the public in several mining papers; but we will merely compare it with the Stetefeldt furnace, and point out the distinctions of the two inventions.

As a cheap chloridizing roasting is a vital question for the industry of silver mining in this country, it is evident that Stetefeldt's discovery far surpasses that of Gerstenhoefer in importance. But the question arises, whether the former, as constructed by Gerstenhoefer, cannot be used as well for chloridizing as desulphurizing roasting? We answer, no. In the Gerstenhoefer furnace only such ores can be successfully treated, which, at a red heat during roasting, have no tendency to sinter or stick together.

But the small particles of a charge of ore mixed with salt are exactly in such a condition while roasting, as to have the greatest possible inclination to sinter and adhere to the shelves. They would thus soon obstruct the whole shaft, and prevent any further work. This has been demonstrated by actual experiments on a working scale.

From the foregoing, it is apparent that the application of the Gerstenhoefer furnace, even for desulphurizing purposes, is very limited, and that certain classes of ore must be entirely excluded from it. This is especially the case with galena ores, which are the most expensive to roast in reverberatory furnaces.

In Stetefeldt's opinion, the shelves in the Gerstenhoefer furnace are perfectly superfluous, and all ores, even galena, can be desulphurized by dropping them through a plain shaft heated by fire-places below, if they are reduced to a sufficient degree of fineness. The escape of unroasted dust from the shaft is of no consequence, as a separate fire-place is constructed for the roasting of these suspended particles in the Stetefeldt furnace. Furthermore, the feeding machinery of the Stetefeldt furnace is based upon a principle entirely differing from that used with the Gerstenhoefer furnace.

That a furnace without shelves is cheaper and easier to construct, more durable, less liable to get out of order, and that it requires less labor and skill to run it, must be conceded by everybody.

Much difficulty was experienced to provide suitable feeding machinery for the Stetefeldt furnace. Gerstenhoefer's apparatus, consisting of fluted rollers, which force the ore through slits in the top of the furnace, would not answer at all. The ore fell down in lumps, and arrived at the bottom of the shaft almost raw. The reason for this behavior is simply the tendency of the particles of all finely-pulverized mineral substances to adhere to each other if a slightly compressed mass of them falls through the air. It is, therefore, necessary to introduce the ore pulp so finely divided, that all the particles can be penetrated by the heat within the short time of their fall through the shaft. To feed the pulp with a blower, as it is done in Keith's desulphurizing furnace, was not considered desirable for the following reasons:

1.   The fall of the ore would be accelerated.

2.   The draft of the fire-places would be impeded by the downward current of the air from the blower.

3.   The formation of dust would be considerably increased.

The feeding machinery in its present shape can be briefly described as follows:

A hollow cast iron frame, kept cool by a small stream of water, rests on top of the furnace. In this frame is inserted a cast iron grate, which is covered by a punched screen of Russia iron, No. 0 for wet crushing, of the trade. Close to the punched screen moves, inside of the hopper, a coarse wire screen, No. 3 of the trade, which is fastened to a frame. The frame has flanches resting upon adjustable friction rollers outside of the hopper, and receives its motion from a crank, with $1\frac{3}{8}$-inch eccentricity. To avoid the motion of a stratum of pulp with the coarse screen, a number of thin iron blades are so arranged across the hopper that their lower edges reach close down to the coarse screen and keep the pulp in place. When the crank is set in motion, the meshes of the coarse wire screen cut through the pulp and drive it through the openings of the punched screen. In this way the ore is introduced in a continuous stream into the furnace. The motion of the crank-shaft was variably tried in Reno, at from thirty to seventy revolutions per minute."

The very satisfactory result of roasting silver ores in Stetefeldt's furnaces at Reno, induced the Manhattan Silver Mining Co. to adopt the same in their mill at Austin, Nevada. Mr. Stetefeldt altered the plan of the Reno furnace somewhat, omitting a dust-chamber between the furnace and the vertical flue. Another important improvement is the application of gas generators (§ 58) in place of the arrangement for the use of firewood. The generators are fed with charcoal. After the proportion of air (which is admitted through a separate flue and is necessary for the burning of the carbonic-oxide gas) has been regulated, the uniform heating of the furnace does not depend on the skill of the fireman. There are three gas generators at the furnace, two communicating with the furnace shaft, a, of Fig. 9, § 58, and one with the vertical flue, b. One separate gas generator could supply all three entrances, but Mr. Stetefeldt prefers to have them separate, partly on account of dispensing with iron gas pipes and other inconveniences connected with a general generator. The choice of firewood or gas arrangement depends on local circumstances—the comparative price of wood and charcoal, etc.

The furnace, as represented in Fig. 9, § 58, is calculated to roast from twenty-five to thirty tons of ore in twenty-four hours, at a cost of from $5 to $6 per ton, while the expenses in usual reverberatory roasting furnaces at Austin amount to $12 or $14 per ton.

Stetefeldt's furnace is of vital importance, especially for poor ores requiring roasting. Considering the temporary repairs of the arch and floor of the reverberatory furnaces, the amount of blacksmith's work on hoes and rakes, besides what was stated above, the superiority of Stetefeldt's furnace over the reverberatory and other mechanical furnaces is obvious.

It may be remarked that wherever the use of the solving process (§ 60) appears admissible on silver ores, this, in connection with Stetefeldt's roasting, will allow the most economical extraction of silver, even from very rebellious ores. The baking of the ore during the chlorination, in the presence of lead, cannot take place in Stetefeldt's furnace, and it is therefore very probable that a higher amount

of lead will be less injurious than in any other
roasting process.

Considering the old, or rather the usual
theory deduced from the roasting process in common reverberatory furnaces, that sulphates must
be formed before the salt can be decomposed, and
not till then will the chlorination begin, it would
seem that for these chemical reactions more time is
required than a few seconds ; but this is not the
case.   As soon as ore and salt enter the furnace,
each sulphuret particle ignites in the glowing at-
mosphere, evolving at the same time sulphur, which
in presence of the oxygen of the atmospheric air,
coming undecomposed through the grates, is
turned into sulphurous acid and the metal into an
oxide, or in part directly into a chloride.   The
sulphurous acid in contact with the ore particles
and oxygen becomes sulphuric acid.   The temperature is nearly from the start too high to permit
the formation of sulphates, so that the sulphuric
acid turns its force on the red hot salt particles,
setting the chlorine free.   All these reactions are
performed instantaneously.   Steam, emanating
from the fuel, is also amongst the gases, consequently the creation of hydrochloric acid must
ensue.   The whole space in the furnace is filled
with glowing gases of chlorine, hydrochloric acid,
sulphurous and sulphuric acid, oxygen, steam, volatile base metal chlorides, etc.—all of them acting,
decomposing and composing, on the sulphurets
with great vigor.   The chlorine decomposes the

sulphurets directly, forming chloride of metals and chloride of sulphur ; it attacks decomposingly also oxides and sulphates, if present. The hydrochloric acid performs the same office. Also metallic silver, if it should occur in the ore, would combine with the chlorine. The sulphuric acid, besides decomposing the salt, oxidizes partly the sulphurets directly, etc.

Considering now an ore particle in a red hot condition attacked simultaneously by all these gases while falling, the final chloridizing result is inevitable. The finer the ore particles are, the more perfect the chlorination ; but even if some coarser parts (to a certain degree) should reach the bottom not thoroughly chloridized, this would be finished in the pile, as the chlorination and evolution of chlorine gas continues in the red hot accumulation on the bottom of the furnace.

38. *Chloridizing Roasting of Silver Ore containing Gold.* Generally the ore containing silver and gold is roasted with salt, converting thereby the silver into a chloride, while the gold remains in a metallic condition. This mode of roasting is quite satisfactory for the subsequent amalgamation in pans. But if those metals are intended to be extracted by a solving process, where no amalgamation takes place, the gold also must be converted into a chloride while roasting. By roasting ores in which gold and silver is present with salt, chloride of gold is formed, according to Plattner ; but before the ore becomes red hot, the gold loses a part of its chlorine, is reduced to a sub-chloride, and, at a little higher degree of heat, to metallic gold.

To form chloride of gold by way of roasting, a better result is obtained in the furnace if the ore is roasted first without salt till the smell of sulphur is no longer perceptible ; and then, after it has cooled down to a low temperature, the salt is added and the whole stirred for some time. A suitable form for a furnace would be a long hearth furnace (§ 49), altered in such a way that the second hearth should be ten or twelve inches below the first ; the arch, however, should continue in a straight line. By this means the space is widened and the temperature brought down to the proper degree. The ore is charged on the first hearth near the bridge, and roasted in the usual way, oxidizing the sulphurets. When this is effected, the ore is shifted over in the lower furnace, and the upper charged again. As soon as the roasted ore assumes a dark red heat, the salt is introduced and raked for two or three hours. According to Roeszner, a combination of gold oxide of soda and chloride of sodium ($Au^2 O^3 Na Cl$) is formed. It is not soluble in water, and but slightly in a salt solution, and cannot be amalgamated, but it is soluble in hyposulphite of soda or lime. V. Lill and others consider the gold in the state of a sub-chloride ($Au Cl$). Hot water cannot be used for the purpose of leaching out base metals, as the chloride of gold would be decomposed.

Proper Roasting—Charges.

39. The endeavor to perform the roasting in the most economical way, leads many operators astray, since they lose sight of the great importance of the chemical reactions, which, as the main object, have to be considered first. Everything has its limit, and so the quantity of ore to be taken for one charge. The European fashion (charging from four to five hundred pounds) could not be well adopted in the United States, while to place 2,000 pounds at once in a furnace might do for a mere desulphurization, but is decidedly too much for both the chloridizing and the oxidizing roasting of silver ores, if good results, by one or the other mode of extraction, are to be obtained. Eight hundred pounds, or at least not over 1,000 pounds, is a permissible quantity in a properly constructed furnace, and with careful handling.

B.—Oxidizing Roasting.

40. The purpose of the oxidizing roasting is either to expel volatile substances which are combined with the metals (as sulphur or arsenic), or to expel volatile metals which are considered obnoxious to further treatment of silver ores (as antimony, lead, zinc, etc). The oxygen has a large share in this transaction, and combines with the volatile substances, as well as with the metals.

Some of the combinations with the oxygen become
volatile — as, for instance, sulphurous, arsenous
and antimonious acids, lead and zinc oxides, etc.
Other combinations again are not volatile, as the
formed sulphates, arsenates and antimonates.
Some of these latter compounds can be disengaged
by an increased heat, as the sulphates of iron and
of copper, whereby the sulphuric acid escapes,
while the remaining metal turns into an oxide.
Others cannot be decomposed by an increased heat,
or an increased heat is considered injurious for
other reasons ; and in this case such combinations
may be decomposed by an addition of charcoal
powder, saw-dust, or the application of hydrogen.
The sulphuric acid is reduced by the carbon to
sulphurous acid, and goes off, and so also the
arsenic and antimony. The carbon deprives the
sulphate or arsenate of a part of its oxygen, and
escapes as carbonic acid.

Changes of the most common Metals while Roasting.

41. *Iron pyrites* and *copper sulphurets* suffer
different changes, based on the action and reaction
of the oxygen, sulphurous acid, sub-oxides and sulphurets, before a complete change is effected into
sulphates of iron and copper. Raising the heat,
the sulphuric acid is driven out, and iron oxide
and copper oxide remain unchanged. If salt is
present, in place of the oxides first, chlorides are

formed, whereby both became volatile. *Galena*, or *sulphuret of lead*, at a low temperature, turns by degrees into oxide and sulphate of lead, the first being volatile. At a higher heat the sulphate remains unchanged and cannot be decomposed into an oxide. In presence of salt the greatest part of the lead becomes a chloride, which is volatile, but a part of it loses some chlorine, and thus being reduced to a basic chloride, is no longer volatile. *Antimony* is volatile as an oxide, but there is also antimonic acid formed, which combines with other metal oxides to antimonates, which are not volatile and not decomposed by increased heat. As a chloride the antimony is very volatile. *Zincblende*, or *sulphuret of zinc*, requires a long roasting. Oxide of zinc is formed besides the sulphate. The sulphate of zinc, under stronger heat, is reduced to a basic sulphate, which can be decomposed to an oxide, but only at a white heat. As a chloride it volatilizes.

What Process Requires Oxidizing Roasting.

42. The oxidizing roasting is in use partly as a preparatory treatment for a chloridizing roasting. It is independent only for Ziervogel's process and for the chlorination process ; that is, so far as concerns the extraction of silver by precipitation. For amalgamation of silver ore no oxidizing roasting is suitable; but it is important with the smelting pro-

cesses, and also in extracting gold from gold ores
—principally from sulphurets (iron pyrites). For
this purpose long furnaces (§ 49) are the most
suitable.

The main point in the roasting for Ziervogel's
process is the creation of a sulphate of silver, and
the oxidation of the base metals as far as possible.
As the decomposition of sulphates of different
metals depends on different degrees of temperature, such roasting appears of a very delicate nature. To this process principally argentiferous
copper matt is subjected.

43. *Roasting of Copper Matt.* When pulverized
until fine enough to pass through a sieve of thirty-
three holes to the running inch, the mass is introduced into the furnace and spread out by means of
rakes. The matt inclines much to clotting. For
this reason a very moderate temperature is applied,
more for drying than for roasting. The matt is
left quiet for about fifteen minutes, after which the
stirring is commenced and continued without stopping for an hour. During this time many lumps
are formed, which the roaster tries to crush to
powder. Near the working door the stuff is exposed to a draft of fresh air, in consequence of
which the roasting on that place progresses more
rapidly than it does further back. This makes a
shifting of the stuff necessary after one hour's
roasting. The other roaster now takes the rake
and stirs the matt again for an hour, doing the
work precisely as the first roaster did. The roast,

ers change in this way every hour for five and one-half to five and three-quarters hours. This roasting is performed on the upper hearth of the double furnace. Twenty-five pounds of coal dust are mixed with the matt, causing an ignition and emission of gases, and the whole mass is transferred to the lower hearth through a hole in the bottom. The upper hearth is now charged with 500 pounds of matt anew.

The sulphur commences to burn after a raking of three-quarters of an hour, and the mass increases in volume when the hearth is covered about four inches with matt. During the roasting all metals are converted into sulphates, of which, toward the end of the operation, iron and zinc vitriol are decomposed, leaving those metals as oxides. Copper, nickel and cobalt remain in the state of sulphates.

The lower hearth is in a light red hot condition when the matt falls in from the upper hearth. To prevent the rapid burning of the admixed coal dust, and the clotting of the mass, a vigorous stirring for an hour, with closed dampers, is strictly observed. The stuff is now shifted and then the damper opened. There now follows a sharp oxidizing roasting, with free access of air, for one hour and a half. By means of the air current, the roasting mass is cooled down so far that it appears quite dark. To see the progress of the roasting, a sample is taken out and examined, either on a porcelain dish or on a filter with cold water. The leach must appear of a clear blue color, and an

addition of salt solution must give some white precipitate, proving the beginning of the formation of
sulphate of silver.   If the filtrate shows a greenish-
blue color, the presence of sulphate of iron is apparent, and in this case the oxidizing roasting must
continue.

The purpose of the addition of coal dust is the
reduction of sulphates to basic salts, whereby sulphurous acid is emitted.   With the opening of
the damper the oxidation progresses, the sulphate
of iron is decomposed almost entirely, the sub-oxide
of copper turns into oxide, and when the oxidizing
roasting is finished, the mass contains mostly
oxides, but also basic salts.   There are copper,
iron and zinc oxides, sulphates of copper and zinc,
while the silver as yet consists principally of an
undecomposed sulphide.   The next stage of roasting at an increased temperature is the last one.   It
is directed toward the sulphurization of the silver
and complete oxidation of the base metals.   It
takes two hours and a half to accomplish this result, under continuous raking and increasing the
temperature to a light red heat.   Samples are
taken again as before, and examined in the same
way.   The leach must appear only of a bluish tint,
and on adding salt solution, a heavy precipitate
must fall, caused by chloride of silver.   The whole
roasting period on the lower hearth, as on the
upper, takes from five and one-half to five and
three-quarters hours.

The formation of the sulphate of silver in the last

period at an increased heat, is due to the sulphuric acid in gaseous form, emanating from the sulphate of copper. It attacks the sulphide of silver, and combines with it to a sulphate. Ninety-two per cent. of the silver is extracted after this roasting. If in the last period the feeding with fuel should be carelessly performed, so as to give a smoky flame, some copper oxide will be reduced to sub-oxide, and this will precipitate metallic silver while leaching, causing a loss. If the roasting should not continue long enough, some sulphide remains undecomposed ; and, on the other hand, if the roasting should last too long, a part of the sulphate of silver would be decomposed to metallic silver and could not be leached out. These circum-stances show that this kind of roasting demands a great deal of attention, in order to obtain a perfect result. The temperature on the lower hearth in the beginning is 500 to 550 degrees Centigrade ; it sinks then to 425, and rises again at the end of the operation to 770 degrees.

44. *Roasting of Gold Ores.* The gold is gen-erally found in a free state as metallic gold. In this state it is easily extracted by proper amalga-mation. Often, however, the gold is combined with other substances, so that amalgamation is of no avail unless the gold is set free by roasting. Iron pyrites and arsenical pyrites are the principal ores containing the gold in a condition unfit for direct amalgamation. Also telluride of gold must

4

be subjected to roasting before amalgamated or chlorinated, but this mineral is not often found.

The roasting of sulphurets and arseniurets is very simple, all that is required being a perfect, dead roasting ; that is, expulsion of all sulphur and arsenic ; but this process takes generally more time than a chloridizing roasting. After the furnace has been heated for some hours, the sulphurets are introduced into it and spread over the roasting hearth, which is generally about twelve feet by twelve, and is capable of receiving one ton. One man is sufficient to attend a single furnace, but a long one requires two men. A single furnace commences with a low heat, sufficient to start the self-burning of the sulphurets, by which so much heat is created as for several hours to require but very little fuel. Nearly half of the sulphur is expelled with this low heat. On exposing a fresh surface of the mass by stirring, the burning of the sulphur with a bluish flame can be seen distinctly. The hoe is principally used for stirring. It must be as light as possible, seven to eight feet long, if prepared to work from both sides of the furnace. The stirring is performed at intervals of ten to fifteen minutes, but not longer ; and wherever the circumstances permit two roasters to be employed, the time of roasting will be shortened. Oxidizing roasting requires more stirring than the chloridizing.

In proportion as the oxidation of the sulphurets draws nearer to the end, the temperature decreases, and it is then necessary to use more fuel to keep the mass at a good red heat. It takes from twenty to forty hours before the roasting of one charge in a single furnace may be considered finished. If, in throwing up sulphurets in the furnace, by means of a shovel or hoe, many brilliant sparks appear, this denotes that the roasting is not finished, but must be continued till this appearance ceases.

In a long furnace (Figs. 7, 8, § 49) the hearth near the bridge is always kept at a bright heat. One man attends to the ore on the first hearth, and the other two or four hearths can be managed by a second. In moving the ore from one hearth to the other, or in drawing the charge from the finishing hearth, these two men assist each other. The finishing hearth receives the ore already desulphurized to a great extent, and containing only a small part of undecomposed sulphurets, but more of sulphates. With a lively heat and active stirring at intervals, all base metals ought to be converted into oxides after ten or twelve hours.

An addition of from thirty to fifty pounds of salt to the ton of sulphurets at the end of roasting, two or three hours before the discharge, is not injurious to the subsequent chlorination, but it increases uselessly the expense if mixed with such sulphurets, when there is no necessity for it. In many instances, however, especially where the sulphurets contain lime, calcspar, talc or heavy spar, a chloridizing roasting is necessary, if it is intended to extract the gold by chlorination. One hundred pounds of salt are sometimes required for one ton of sulphurets.

Sulphurets containing gold can be brought into a soluble condition by means of roasting, according to § 38, so that no chlorination is required after roasting.

Roasting Furnaces.

45. Roasting not only requires much care, but it is also an expensive operation. For this reason the choice of the right kind of furnaces is of very great importance, and so much the more as a perfect and economical extraction of silver depends principally on the result of roasting. The chloridizing roasting is known to be the most suitable way for the subsequent extraction of silver in whatever way it may be performed, by amalgamation or solving ; consequently those furnaces in which the ore particles are exposed to the action of chlorine and other chloridizing gases to the most advantage, must be considered the best. The old style of furnace was four to six feet wide and ten feet long, and in them a small part of the ore was exposed to the greatest heat near the bridge. The gases evolved were carried along by the draft, being in contact with the surface of the ore for a length of ten feet while passing over it ; but on account of the narrowness of the hearth, the ore at the bridge had to be changed often with the cooler part at the flue.

The next step in improvement was the adoption of wider hearths, even wider than long. The heat was more uniform and the result better. In both kinds of furnaces the chlorination of the metals depends principally on the chlorine developed in the mass of the ore while passing through it ; but once above the surface, the chlorine and volatile chloride metals have little chance to transmit their chlorine to the ore (§ 23), and this only through the chlorination period. During two or three hours of each charge, when desulphurization and sulphatization are going on, this must be performed by the oxygen of the air, while, if chlorides were present from the beginning, sulphurets, sulphates and oxides would have been partly decomposed directly by the chlorine, whereby time and a certain percentage of salt are saved.

In this respect a great advantage is gained by the introduction of " long furnaces " (§ 49), in which a continual formation of chlorides on the finishing hearth near the bridge is going on, volatile chlorides and free chlorine being evolved, which, on their way to the flue, are constantly in contact with the ore for a space of thirty or fifty feet in length. These furnaces show a great economy in fuel, labor and salt, and the roasted ore contains a better percentage of chloride of silver (§ 36).

Another most important improvement in the way of chloridizing roasting is found in the Stetefeldt furnace (§ 58), where all ore particles are involved

in chloridizing gases under very favorable circumstances.   The roasting is cheap, and from twenty
to twenty-five tons of ore are roasted in twenty-
four hours—more than ever accomplished in any
other furnace.

The roasting furnaces do not require a white heat;
hence common bricks can be used ; but it is nevertheless advantageous if the fire-place above the
grates is built of fire bricks.   In new or unpopu-
lated districts even unburned bricks or adobe may
be used ; they stand just as well as burned bricks
of the same material, except on the floor of the furnace, which is worked out in two or three months.
Hard bricks are the best material for the hearth-
floor, placed edgeways (four-inch), with as little
clay between as possible, and laid carefully and
well fitting, so as to form a level and smooth surface.   All parts exposed to heat must be built with
loam or clay, not with mortar.   Many masons have
the custom of laying three heights of bricks so that
the eight-inch wall is formed by two rows lengthways, and only the fourth height is put crossways.
It is a quick work and may answer for buildings,
but should not be allowed with furnaces where the
expansive heat must be considered, especially in
the fire-place.   Each alternate row of bricks must
be laid crossways to the preceding ; also, adjusting
the wall with the hammer, to make it perpendicular
and square, after several bricks are laid, is injurious.
The outside appearance of a furnace is of minor importance, and the mason must, contrary to his gen-

eral idea, pay the most attention to the solid and particular work inside. The distance of the arch from the hearth is from eighteen to twenty inches in the highest point, not far from the bridge ; in a long furnace, however, the roof of the first hearth can be higher from the floor by four to five inches, according to the length. An eight-inch thickness of the arch is sufficient, and the bricks laid with the eight-inch side perpendicular form a more durable arch than one of twelve inch thickness composed of eight and four-inch sides of the bricks. The furnace must be secured against expansion by grappling-irons of cast iron tightened with iron rods from five to six-eighths of an inch in diameter. The rods placed over the length of the furnace are stronger—one inch in diameter. In place of iron grapplings, also wooden posts, six by eight inches, are used, tied by iron rods on the top. The lower ends are generally put in the ground, but it is preferable to use rods on both ends. In case of need, even the rods are replaced by timber. For the passage of the rods square holes must be provided. in the masonry; also for the escape of dampness such passages are necessary at different points, especially if the whole block consists of masonry. The floor of the hearth should be three feet and a half above the ground ; if lower, it is inconvenient for the roaster.

There are two principal classes of furnaces—such as are managed by hand and such as employ machinery. For the first class mostly reverberatory

furnaces are in use.   The second class comprises
reverberatory, cylindrical and vertical furnaces.

A.   Roasting Furnaces Managed by Handwork.

46.   *Reverberatory Furnaces.*   Reverberatory
furnace is the name applied to all horizontal hearth
furnaces provided with grates and fire-place on one
side, and a flue connected with a chimney on the
other.   The draft here is created by the chimney
instead of by bellows, as in blast furnaces ; therefore only such fuel is used which gives a flame, and
consequently no charcoal, coke or anthracite is
serviceable unless in a gas reverberatory furnace,
where gas (carbonic oxide) is produced from charcoal or other fuel—sometimes also by the aid of
compressed air—and burned.   The reverberatory
roasting furnaces are constructed in various ways.
There are single furnaces, with but one hearth, and
double furnaces, with two hearths, one above the
other.   Sometimes above the second hearth there
is a third one for the purpose of drying the charge.
Long furnaces also are coming into use.

47.   *A Single Roasting Furnace* is represented by
Fig. 3, showing the section, and Fig. 4, the ground
plan.   The bottom, *a*, or the hearth, is made of
the hardest bricks, laid edgewise and as close as
possible.   Some masons lay the bricks flat.   This
mode is cheaper and quicker, but far inferior and

less durable than the former way, and requires a more carefully prepared foundation. The very

Fig. 3.

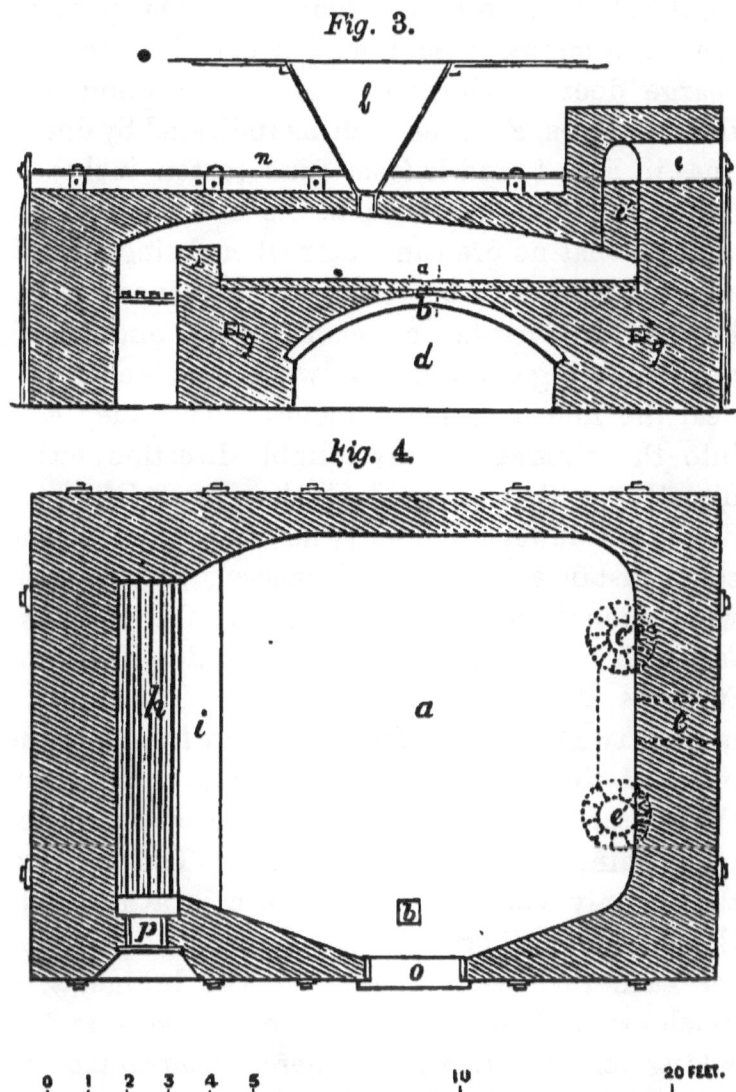

Fig. 4.

best bricks must be selected for the hearth. *b*

4*

shows the discharge hole in front of the hearth. It is more convenient to draw the ore toward the front hole than to have a door for this purpose behind, but circumstances may decide for such discharge doors. The flue, e, is in connection with the flue-holes, e', in the arch, as indicated by dotted lines in Fig. 4, and is from nine to ten inches in diameter. The flue-holes in the arch have the advantage that no ore can enter when being stirred, as often happens when the flue commences at the hearth. The distance between arch and hearth near the bridge is twenty to twenty-one inches, and near the flue only eight inches. The flue leads into the chimney in any suitable direction, either directly or through a dust chamber. Often the flue is led under the floor (when the chimney is at some distance from the furnaces), and is made wide enough to serve as a dust chamber—say two feet wide and three feet high, or wider if several furnaces are connected therewith. The chimney is from twenty to fifty feet high, and from one and one-half to three or four feet square in the clear. On the top of the chimney an iron cover, controlled by a chain, regulates the draught. This is practicable only when but one furnace is attached to the chimney, otherwise dampers must be provided for each furnace in the flue. The bridge, i, is much exposed to injury by fire on one side, and by raking on the other; it is therefore advantageous if the upper part, or the whole bridge, can be made in two or three parts and of some fire-proof stone,

—sandstone, granite, or some conglomerate, which does not burst when heated. The grates, h, are twelve to sixteen inches below the top of the bridge, eighteen inches wide, and from six to seven feet long. The space between the grate-bars is one-fourth to one-half of an inch.

In the roof, nearer to the bridge, is an opening four to five inches square, of cast iron, in connection with a funnel, l, of sheet iron. This funnel must be large enough to receive one charge of the ore. A slide keeps the ore in the funnel. The roof must be either eight inches thick, or the double length of brick; that is, sixteen inches. Under the hearth there is an arched space, d, into which the roasted ore is drawn through the discharge hole, b, either directly into an iron car or on an inclined floor, on which the ore slides from underneath the furnace. In front this space is shut up by brickwork. For the purpose of easy drying it is well to leave open some holes, g, for the escape of dampness. It is not necessary to build the block under the hearth solidly of bricks. The space inside is generally filled up with rubbish of bricks and stone.

The working door, o, is from twenty-five to thirty inches wide. In front of it is an iron roller for easier handling of the heavy tools. The door is eight to nine inches high. The cast iron door-frame, p, for the fire-place, is from nine to twelve inches square. When completed, the furnace is tied by iron rods, n, both ways. The uprights are generally wooden ones, six by six or five by eight inches.

It is very important to dry the furnaces, when finished, with a very moderate fire for five or six days, day and night. Upon a slow, gradual drying, the durability of the arch depends. The furnace must be nearly red hot before the first charge of ore is introduced.

48. *A Double Roasting Furnace* is represented

Fig. 5.

in Fig. 5, in longitudinal cross section. The lower hearth, *a*, is nine feet long and ten feet wide. The

roof in the center is eighteen inches, and at the
flue and bridge fourteen inches above the hearth.
The fire-place, r, is twenty inches wide, eight feet
long, and twenty inches from the roof. The flue,
b, ascends to the upper hearth, c, the working door,
o, of which is on the back side. In case there should
be required more heat than is obtainable from
the lower hearth, there is an auxiliary fire-place, r'.
The flame goes through the flue, b', into the dust
chamber, g. This chamber has four cross partitions lengthways, by which the draught is forced
to take a longer way before it enters the chimney.
From the upper hearth the ore is drawn through
the hole, d, to the lower hearth, as the bridge does
not permit the flue to be used for this purpose.
e, e, are canals for the escape of moisture. The two
hearths can be used separately if needed. In this
case the flue, b, is closed and another one (not seen
in the drawing) opened. This second flue communicates directly with the dust chamber.

49. *Long Roasting Furnace.* This kind of
roasting furnace, as represented by Fig. 6 in vertical
section, and Fig. 7 in ground plan, gives much
satisfaction, as there is not only a great saving of
fuel effected, but also a greater quantity of ore can
be roasted in a given time than with a single furnace. It is only a modification of the double
furnace, but it seems to be more convenient for
the roasters. The heat is better utilized, as the flame
has not to pass through flues between the hearths,

and is not broken so often, but the moving of ore from one hearth to the other is more troublesome. There are two men employed at a time, there being one ton and a half to two tons in the furnace. The hearths are either arranged horizontally, as the drawings show, or only the first one is level; the other two are inclined ; this facilitates the shifting of the ore. Each hearth is ten feet long and ten or twelve feet wide. After the first hearth there is a step of four to six inches, partly to divide the first from the others, but principally to contract the space between roof and hearth of the other two. The ore is fed on the last hearth through the sheet iron funnel, *a*, spread equally on *b'*, and, according to its dampness or the quantity of sulphurets contained, stirred more or less for one and a half to two hours. As it is not only inconven-

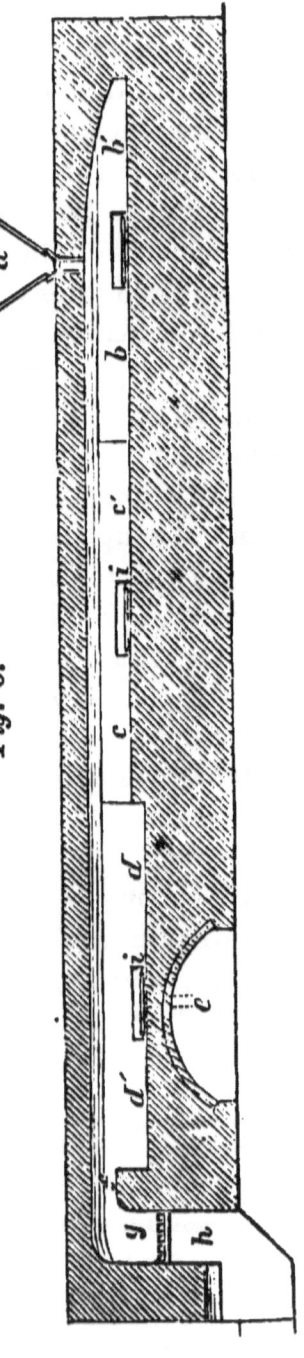

Fig. 6.

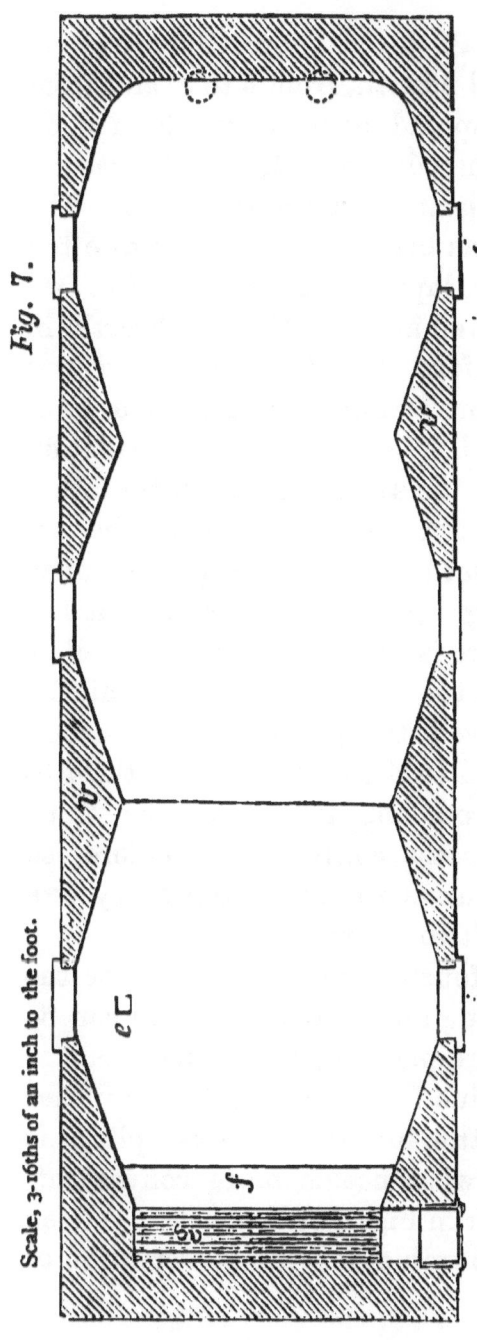

Fig. 7.

Scale, 3-16ths of an inch to the foot.

venient, but impossible to have a good stirring effected at a distance of twelve feet, which requires long and heavy tools, there are for this reason working doors on both sides of the furnace. The roaster uses hoes or rakes eight feet long, made partly of gas pipe, which are light and handy. The working doors are thirty inches wide. They must all be kept closed except when the ore is being raked, and then it is very proper to have half of the door closed (with a piece of sheet iron). Sufficient air comes in at the working door of the first hearth.

After one and a half to two hours the ore is removed to the second hearth, from b to c' and from
b' to c, and again spread over the whole of c, c'.
Another charge is introduced on b, b'.   The second
hearth has a better heat than the third one.   The
ore is treated here as before, being raked as often
as possible.   After a lapse of one and a half to two
hours the ore is moved again to the first hearth, in
the same way as before ; that is, from c to d' and
from c' to d.   The ore is now exposed to a light
red heat, by which the chlorination or oxidation
must be finished in the same time as on the other
hearth.   It is necessary to change here the ore
from the bridge toward the flue, and reverse once
during the. roasting.   When the operation is finished, the roasted ore is drawn into iron cars below
the furnace through the opening, e.   When all the
ore has been removed, the charge on the second
hearth is transferred to the first, from the third to
the second, and from the funnel to the third
hearth, and the process continued as before, so
that a thousand pounds are drawn out every one
and a half or every two hours.

The bridge, f, is fourteen inches high.   For the
purpose of admitting air or steam, a canal can be
made in it.   The fire-place, g, is eighteen inches
wide and eight to nine feet long, and fifteen inches
below the top of the bridge.   The ash-pit, h, is
made according to what seems more convenient,
as represented either in Fig. 6 or in Fig. 3.   A deep
ash-pit is more favorable for the preservation of

the grates, as they are less heated. Each door is provided with an iron roller, i. A furnace of a similar description is in operation in La Dura (Mexico), roasting refractory silver ores for the chlorination process.

A furnace sixty feet in length, with six hearths, as built by Mr. Graff at the San Marcial, has the advantage of being capable of roasting from eight and one-half to twelve tons of ore in twenty-four hours, discharging every hour from eight hundred to one thousand pounds, according to the charge. In case ore is subjected to roasting which has not enough sulphur to create the required heat in burning, an additional smaller fire-place must be attached on one side, so as to bring the flame into the fourth hearth.

Muffle Furnaces.

50. A muffle furnace, as the name indicates, is a furnace constructed of clay and cast iron in such a way as to prevent the flame from coming inside of it. The fuel heats the mantle or muffle from the outside, so that the ore is not heated directly by the burning fuel, but by the glowing muffle. The muffle furnaces require, therefore, more fuel to obtain a certain degree of heat than ordinary reverberatory furnaces, where the flame comes into contact directly with the ore. The use of this furnace is limited, and applicable in cases where the air or the gases of the burning fuel are injurious, or where volatile substances from the ore should be condensed ; as for instance, sulphur, zinc, arsenic, etc. .For roasting silver ores, these furnaces are not in use, but they were tried in California in different ways ; also for desulphurization, adding charcoal to the pulverized ore. The experiments, however, were not successful, as could have been anticipated (§ 5, e).

B. Roasting Furnaces with Mechanical Arrangements.

51. There is a great variety of furnaces wherein the costly stirring by hand is replaced by mechanical apparatus. No mechanical furnace can be governed in every part of the roasting process with the same facility and precision as is possible in a reverberatory furnace with manual labor ; but in the latter case the great difficulty in finding good reliable roasters, and the heavy expenses connected therewith, make a mechanical substitution very desirable. In one respect, vertically revolving furnaces, in which the ore moves without being stirred by mechanical rakes or plows, have the advantage, viz : in simplicity. The stirring furnaces combine sometimes complicated machinery with a general defect, and this is the wear of the shoes or plows, not on the lower part alone, but also on the sides. Some stirrers are fixed, and consequently cannot remedy the wear by sinking ; but none have been yet so constructed as to keep their

original line on the wall side. The consequence
is, that while the shoe wears away, the ore takes
its place and hardens there (being exposed to heat
so long) till new shoes are put in; and in this case
all the hard cakes are broken off and mixed with
the ore. But as new shoes are not put in every
day, these periodical lumps may be sifted and re-
turned to the battery. The accumulation of ore in
some parts of the furnace successively, cannot be
avoided entirely, even in furnaces where no stirring
goes on.

A. Stirring Furnaces.

52. *Revolving Hearth Furnace.* The shape of
this furnace is circular. There is an iron frame of
from ten to twelve feet diameter, with sides fourteen inches high. The whole is lined with brick,
the bottom four inches thick. The discharge opening is on the bottom, extending from the periphery
toward the center, and is four inches wide and
three feet three inches long. This opening is shut
by an iron door, hung on hinges. It is not necessary to fill this space with brick, which would interfere with the easy opening ; but the space, after
the discharge of the ore, must be filled up with
roasted ore, of which enough is always left in the
furnace. The bottom is fixed to an upright shaft,
four inches in diameter, provided with a spurwheel
at the lower end to impart the motion. This ten
or twelve-foot bottom is surrounded by a substan-

tial ring wall, as close to the periphery of the bottom as possible. The bottom is then arched over with bricks, leaving the doors through which the new shoes are introduced when the old ones wear out. There is also a cast iron pipe through the center of the furnace, on which the shoes are fastened and so arranged that one set plows the ore against the center, the other set toward the periphery. The pipe is hollow and cooled by a continual stream of water. There is also a hole four to five inches square in the arch, in connection with a funnel, through which the ore is charged into the furnace. The distance from the bottom to the center of the arch is thirty-one inches. The arch is connected on one side with the fire-place, six or seven feet long and eighteen or twenty inches wide, and about ten inches below the rim of the revolving hearth are the grates. On the opposite side is the connection with the flue.

Such furnaces have the advantage that they carry the ore in a circle, so that each part is equally exposed to the heat near the bridge and to the cooler region near the flue. While revolving, the funnel is opened and the ore falls on the moving bottom, being spread in passing under the stationary stirrers, which are of a plow shape. The roasting takes about the same time as in an ordinary furnace, but requires less fuel, as the furnace is not cooled down by air, which enters the common reverberatory furnace through the working door. It is important to have the horizontal shaft provided with two driving wheels of different size, so that about one to two revolutions per minute can be obtained while roasting, and from six to eight revolutions while discharging. After the roasting is finished, the discharge door on the bottom can be opened, while the hearth revolves slowly. In this furnace it is an easy matter to arrange the plows in such a way that they could be moved every second or third day toward the periphery as much as they wear off. In this way the side of the hearth can be kept always clear from accumulation of the ore crust.

53. A similar furnace is Brunton's revolving furnace. The hearth has a low conical shape, the highest point being the center. Above this is the charging hole in the roof. The hearth is twelve feet in diameter, and takes one ton of tin ore at a charge. There is a cast iron rake with three-inch long prismatic teeth, which are dovetailed and so constructed as to be easily replaced. The ore comes through a funnel in the center of the revolving hearth, and is spread by the stationary rake, the position of which is not radial but oblique. The hearth is fixed to a solid vertical shaft with gearing, by which a slow rotating motion is imparted to the hearth, so that only one revolution is made in forty minutes.

54. *Ernst's Rotary Furnace* is constructed on the same principle as the former two, but differs

materially in two points. While in the before described revolving hearth furnaces, the diameter cannot be increased well beyond twelve or fifteen feet,
principally for the reason that the roof would stand
too far off from the bottom, Ernst's furnace can
be constructed on any given diameter. The other
point of difference lies in the discharge, which is
continuous with Ernst's furnace.

The hearth is a circular iron ring (disk), lined
with bricks, and revolving on a series of iron rollers. It is kept in motion by means of two gear
wheels, each three feet in diameter. The speed is
regulated by a cone pulley, being increased or diminished to conform to the required motion of the
hearth. The ore is charged continuously from the
battery, through a hopper, by means of elevators.
The hearth moves constantly, carrying the ore from
the flue toward the fire-place, and exposing it to
all the different temperatures of the furnace, thus
effecting a very uniform heating of the ore. The
hearth-ring is surrounded by a corresponding wall
of masonry, leaving a free circular space in the
center, where the driving machinery is placed.
There are stationary and movable stirrers. The
movable paddle-stirrers are connected with the
gearing under the iron bed of the hearth on its inside periphery. The ore finally arrives at the fire-
bridge, where it receives its ultimate and highest
heat, and is then discharged by an apparatus,
which consists of two chain or rag wheels, by
which a double endless chain, with inserted plates

of iron, is carried across the hearth, thus discharging the roasted ore.

This furnace, not as yet in use, promises to do
good work. In some respects it could be compared with O'Hara's furnace—having a long, nar-
row hearth ; replacing the endless chain, however,
by the motion of the hearth, both being self-dis-
charging. For chloridizing roasting of silver ore,
the center circle of the ring must have a diameter
of about twenty-six feet. One revolution should
be accomplished in five or six hours. It takes
about twenty stirrers to rake the ore every ten
minutes, if equally divided. It is, however, more
proper to arrange the stirrers closer toward the
feeding place.

55. *Parke's Roasting Furnace*, with movable
stirrers. This is a double furnace, one hearth
above the other, with a common vertical shaft to
which the stirrers are fastened. The hearth is
twelve feet in diameter and rests on an arch, beneath which the rotating motion is transferred to
the shaft by means of gearing. On one side of the
lower hearth is the fire-place, whence the flame
draws over the bridge into the furnace.

Opposite the bridge is an opening one foot wide
and four feet long, through which the flame ascends
to the upper hearth. Both of the hearths have
two working openings, which are closed by cast-
iron doors. From the upper hearth the flame
draws through a flue into the chimney. The shaft

goes through both hearths and the roof. There are two massive arms in both furnaces, with curved spikes attached for the purpose of stirring the ore. In order to keep the shaft cool, it is hollow, and a few holes above the gear permit the cold air to draw through the shaft, whereby a constant cooling is effected. The upper end of the shaft runs in a cast iron cross, fixed on the roof of the furnace.

After the ore on the lower hearth is drawn out through the discharge hole at the bottom, the ore on the upper hearth, already desulphurized to a great extent, is raked toward a similar discharge hole, and then transferred to the lower department. The raw ore is charged through the roof into the upper part. By means of hoes the ore is spread on both hearths, before the shaft is allowed to revolve again.

56. *Bruckner's Revolving Furnace.* An iron cylinder about ten feet long and four feet in diameter is lined with bricks inside. The cylinder is fixed with its long axis in a horizontal position. One end communicates with a fire-place, from where the flame passes through the revolving cylinder into the flue at the other end. Inside there is an inclined shelf, the position of which is such that the ore is being continually shifted from one end of the cylinder to the other, as this revolves and thereby thoroughly exposed to the flames. In this way the ore becomes uniformly heated. To obtain a satisfactory result, the furnace must revolve very slowly. It makes one revolution in two to five minutes. The ore is carried up by the revolving furnace to a certain height, whence it falls through the flame. The draught through the cylinder carries a part of the finest ore out through the flue ; it is therefore necessary to build dust-chambers between the flue and the chimney. On the long side of the cylinder is an opening, closed by a lined iron door, through which the ore is discharged when finished. Through this same door the ore is charged. This kind of furnace has been in use at the La Dura works for several years, where it has given satisfaction. It requires less fuel and less labor than the single reverberatory, and only of late has it been replaced by a long furnace. Some are still in use in Colorado Territory.

57. *O'Hara's Mechanical Chain Furnace.* Of all furnaces, the object of which is a continual discharge of roasted ore, taken directly from the stamp without the intervention of manual labor, O'Hara's was the only one crowned with practical success. Stetefeldt's furnace of late is arranged in the same way—that is, concerning the direct feeding from the stamps by machinery. O'Hara's furnace is in use on Carson River, Nevada. Three of these were in operation at Flint, Idaho Territory. At present, however, they have stopped for want of ore. The construction of O'Hara's furnace is shown by an outline drawing, as represented in

5

Fig. 8.　The hearth, A, is 104 feet long and nearly five feet wide. Eighty feet of this hearth are crossed by an arch, B, twelve inches high, and connected with three fire-places, two, c and d, on one side, and one between c and d on the other. a is the feeding hearth, provided with ore continuously from the batteries. The motion of the ore is effected by an endless chain, g, passing over two chain wheels, one at each end. To this chain two oblong flat rings, h, are attached, each provided with eight shovels or plows so arranged that while one of the rings shoves the ore toward the center line, the other pushes it back again toward the sides every three or four minutes (or in shorter intervals if more ore is charged). The ore not only changes its place to the right and left, but it also moves forward by degrees, so that in the course of six hours from the beginning, it commences to be discharged at f, passing eighteen feet over the cooling hearth, e. On both ends of the furnace are iron doors hung on hinges, which are opened by the rings. After several months of operation the hearth or bottom appeared in good condition.

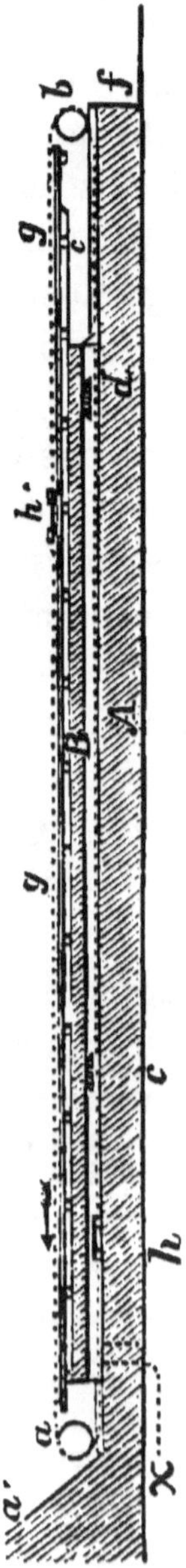

Fig. 8.

The five batteries, five stamps each, have on both long sides endless screws, by which the crushed ore is forwarded, in proportion as it is discharged, to an elevating apparatus. Being lifted about fifteen feet, it is conveyed again by endless screws along the feeding hearths of all three furnaces, a', and regularly divided and discharged on the feeding hearth, a. The ore, mixed with 100 pounds of salt to each ton, is spread on iron plates before the batteries, (heated by the hot air from the furnaces, conveyed through the flue and under the plates.) When charged into the battery the ore is not further handled till it comes out of the furnaces perfectly roasted (§ 32).

There is only one obstacle connected with this and other mechanical furnaces. The shoes or shovels, touching the sides of the furnace, wear off by degrees, leaving a space which is taken up by the ore. This part of the ore along the wall hardens and increases in amount in the furnace till new shoes are put in. By these the crust of one-half to three-quarters of an inch thick is broken off and carried out. From the Rising Star ore these crusts contain nearly just as much chloride of silver as the well roasted ore ; they are, nevertheless, disagreeable, but some means might be devised by which this inconvenience could be avoided.

58. *Stetefeldt's Roasting Furnace.* This furnace, now being built at Austin, Nevada, is represented in Fig. 9, showing a vertical cross section. The furnace at Reno, Nevada, has a dust-chamber in place of the flue, *b*, of Fig. 9, the omission of

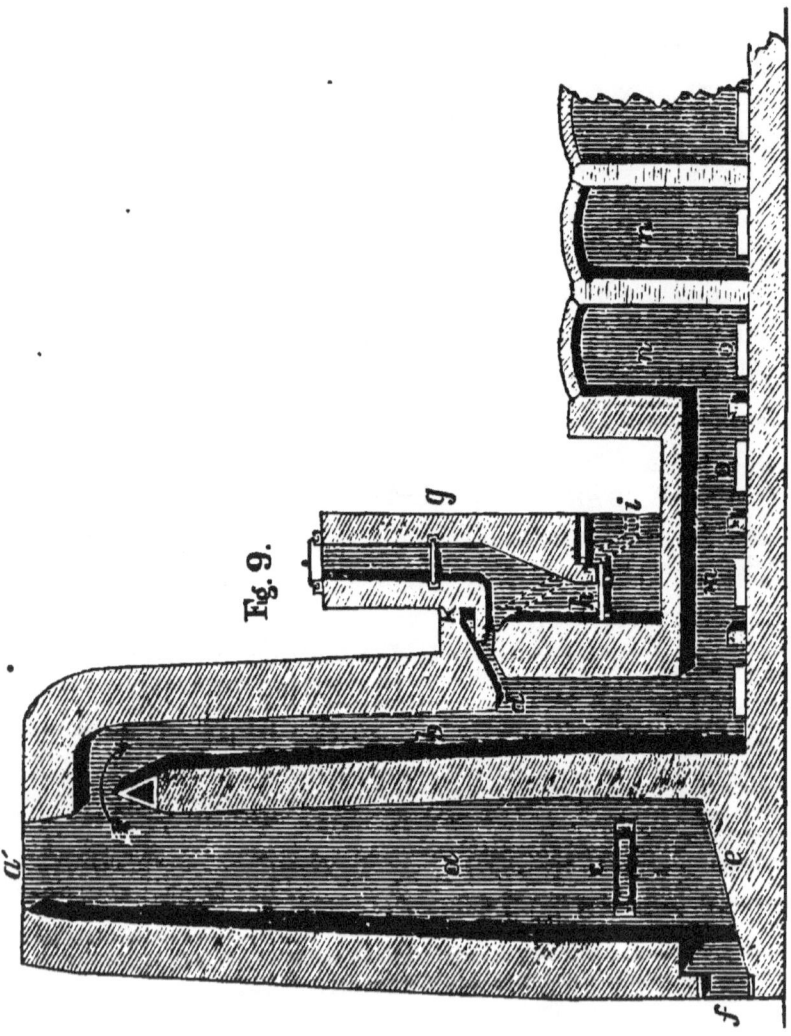

which simplifies the construction without injury to the good results of roasting.   The furnace has three important departments.   1st. The roasting shaft, *a*, twenty-five feet high and five feet wide at the bottom, narrowing somewhat toward the top, to prevent the adherence of dust to the wall. It is a simple shaft of common bricks, built as smooth as possible. On the top of the shaft, at *a'*, is placed an iron feeder, through which a permanent and uniform feeding of the pulverized ore, already mixed with salt, is effected. The ore falls on the bottom, *e*, and when half a ton or a ton is accumulated, it is drawn out through the door, *f*. 2d. The fire-places. There are three gas generators, constructed similarly to that of the copper-refining furnace at Mansfield, Prussia. The cover is taken off and the charcoal introduced. The cover is placed again on its frame, which contains sand in a groove in order to shut off the draft entirely. The slide door near *g* is drawn out, and the charcoal falls on the grate, *h*, through which as much air is admitted as is necessary. There are also two canals on each side of the grate, one of which is shown by dotted lines, *i*, both communicating at *k*. Through these canals is regulated the admission of the air for oxidizing or burning the carbonic acid, created above the grate, *h*. In the flue, *d*, air and gas meet together, and the burning product heats the furnace. Two of these generators heat the shaft, *a ;* the mouth of one is shown in the drawing by *c*, the other is on the opposite side, and therefore not visible in the plan. The two generators are constructed exactly like *g*, with the exception that the flue, *d*, is not inclined, but hori-

zontal. The flue, d, as well as the generators above
the grates, are lined with fire-bricks. 3d. The
dust-chambers. With the draft, the gases from
the shaft, with a part of the fine ore dust, pass
through the vertical flue, b, then through the horizontal one, m, into a series of chambers, n, of different sizes. The first four chambers, n, are
smaller than the four following, which are not represented in the diagram ; from the last chamber
the gases draw into the chimney. The dust can be
removed from the bottom of the chambers through
the doors, o, o. Almost all the dust is regained,
and not in a raw condition, as from dust-chambers
of reverberatory furnaces, requiring re-roasting,
but perfectly chloridized, which is principally due
to the auxiliary generator, g, and the longer contact with the chlorine gases.

Chimneys and Flues.

59. The draft in a furnace depends on the
height of the chimney. The flue or canal between
the hearth and chimney has a great influence on
the draft, as a great deal of heat is taken up by the
walls, and the draft in the chimney depends on the
temperature therein up to a certain degree. It
follows that the longer the flues are, the higher the
chimney should be. Flues underground, once
heated, absorb less heat than if exposed to the air.

Single roasting furnaces, each having its own
chimney, dispense entirely with long flues. It is

therefore sufficient to build the chimney twenty to twenty-five feet above the level of the hearth, and fifteen to eighteen inches square in the clear.

Underground flues are suitable where many roasting furnaces are connected with the chimney. They are often built directly under the furnaces, two feet wide by three to four feet high. In this position the connections between the main flue and those of each furnace are the shortest. Although this mode is preferable to flues on the side of the furnaces, it is not practicable where deep ash-pits are in use. Deep ash-pits are favorable, partly for the reason that it is not necessary to carry out the ashes every day or two, but principally on account of the grates, which last longer, being better cooled by the air.

The flues are sometimes needlessly carried too far out, especially when it is intended to reach ascending ground on which the flue continues, replacing the chimney entirely or in part. The ascent must be steep, otherwise if the length is in no proportion to the perpendicular height gained by it, taking also the distance from the furnaces to the ascending ground in consideration, the absorption of heat might neutralize the advantage of the ascending flue. On the other hand, the temperature in the chimney should not exceed 300° C. = 572° F.

A chimney fifty to seventy feet high, and from three to four feet square inside, is generally considered sufficient for a number of roasting fur-

naces.   It is built of common bricks, sometimes of
stone.   In the latter case the stone must be examined to ascertain whether it will stand the heat
without bursting.   The stone work is often lined
inside with a layer of bricks.

According to experience, the most advantageous
proportion between the area of the grate-openings
and the section of a chimney, is between one to
one and two to one.   The round section is the
most proper, as there is the least friction with it.
The section ought to be the same at all heights.

III.   EXTRACTION OF SILVER

BY WAY OF LIXIVIATION.

Solving Process.

60.   Under " Solving Process " is to be understood here, simply roasting with salt, and extracting
the silver with hyposulphite of lime or of soda,
without reference to particulars of roasting in
the Patera or Kiss processes.

The solving process comprehends, generally considered, different modes of extraction, all of which

are based on the property of the solubility of the chloride and sulphate of silver. The extraction of the chloride of silver by alkaline hyposulphites was proposed by Percy. Patera was the first who made use of the hyposulphite of soda for extraction of silver in a practical way; his success, however, depends principally on his modified and complicated roasting (§ 33). By lixiviation the silver is extracted in the Patera, Kiss, Röszner-Patera, Ziervogel, Augustin, and Kustel & Hofman processes.

The extraction of silver by the solving process is simple. The ore is first roasted with salt in the usual way, whereby the formation of base metal chlorides cannot be avoided entirely. After roasting, the ore is first subjected to leaching with water, in order to extract the base metal chlorides, and then with hyposulphite of lime, to extract the silver.

The Extraction of Silver.

61. After a chloridizing roasting the ore should be examined to ascertain the amount of chloride of silver contained in it, according to § 21. In case the extraction should not be satisfactory, it is then easier to find what the cause is. The ore is then prepared for leaching.

A. First Leaching. The roasted ore contains chloride of silver, which does not dissolve in water, but generally there are also base chlorides in it, as the chlorides of copper, zinc, lead, iron, antimony,

5*

etc., which are soluble. It is the purpose of the first leaching to extract these base metals by means of hot water. For this purpose the ore is introduced into a tub or square box of pine wood, the planks being one and one-half to two inches thick.

Fig. 10.

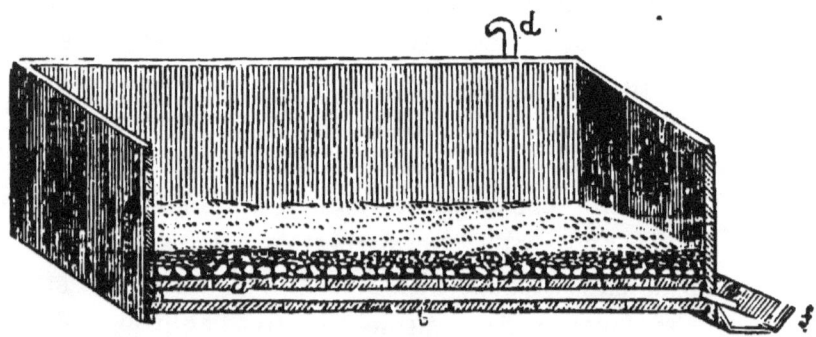

The boxes must be made as water-tight as possible and provided with a filter at the bottom. The filter is prepared in two ways, either as represented in Fig. 10 by fixing a false bottom, a, provided

Fig. 11.

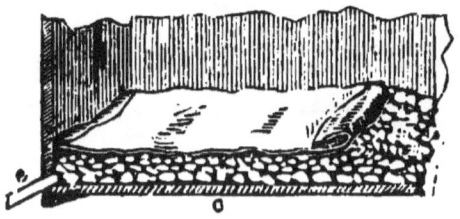

with numerous holes, one-half inch in diameter, about one inch above the bottom, b, or as Fig. 11 shows, without a false bottom. On the bottom, a, is thrown clean rock, quartz or poor ore of about the size of a hen's egg, three or four inches high ; on this smaller stuff, and finally sand, free from mud. In Fig. 11, rock of about the same size is thrown directly on the bottom, c, spread four inches high, then a few buckets full of rock, not smaller than hazel nûts, and only so much of it taken as to equalize the surface. This is then covered with a piece of canvas and is ready for use. The boxes, according to their size and the weight of the ore, may contain from one to five tons. Generally the ore must not be over fourteen or sixteen inches deep, but some ore allows a good leaching with twenty-five inches.

The roasted ore, generally without sifting, is charged into the boxes, and the surface spread evenly, leaving about six inches space from the top for the reception of the leaching water. The hotter the water is, the sooner it dissolves the soluble salts and the quicker the leaching progresses. It is conveyed through the pipe, d, and falls on a piece of canvas, whence it spreads equally and gently over the ore. The water soon reaches the bottom and begins to flow out through the pipe, e, into the trough, f.

In the beginning, the leaching water at e is highly charged with base metal salts, and shows a green color if there is much copper in the ore. The water is kept running in a continual stream till it reaches nearly to the rim of the box, when the influx and the efflux are equalized. After one or two hours a glass full of the liquid, at the pipe, e, is taken, and a few drops of sulphide of calcium

(or of sodium) added. If a precipitate falls, of a
dark or light color, the leaching must continue ;
but it is not necessary to continue until no precipitate at all is perceived, as it requires some time—
perhaps an hour—before all the water runs out
after the pipe, *d*, is closed. The water which
comes out last must be free from salts. This first
leaching takes from two to four hours, sometimes
longer.

B. Second Leaching. As soon as the ore is
freed from the base chlorides soluble in water, a
solution of hyposulphite of lime (§ 70) is led in
from a tub or tank, on the ore, in order to dissolve
the chloride of silver. This leaching is conducted
like the former. It depends on the amount of
silver how long this work continues—from eight to
twenty hours. The clear cold solution, containing
the chloride of silver in the form of a double salt,
has a very sweet taste, and is conveyed through a
trough or india rubber hose into a precipitating
tub. Very rich ore, containing 12 to 15 per cent.
of silver, would require forty-eight hours leaching,
and even then it would be necessary to subject the
ore to a second leaching with the hyposulphite,
with an intermediate roasting with green vitriol
and salt ; for, with the best work, if 95 per cent.
are extracted, the tailings would still appear sufficiently rich for this, containing about 200 ounces
of silver per ton. Ores containing $350 per ton
are often leached out perfectly in twelve hours.
The end of the lixiviation is ascertained in the

same way as in leaching with water, using the sulphide of calcium. If no precipitate is obtained the extraction is finished.

The color of the precipitate is a black-brown. The presence of base metals changes the color somewhat. Iron makes it black; copper, red-brown; lead and antimony, light red-brown, etc. The silver is first dissolved, especially if a diluted solution of hyposulphite of lime is used; and for this reason the first precipitate is the richest in silver. Ore containing a great deal of lead—especially if the roasting was so conducted that a large part of it remained as sulphate of lead, which is not soluble in the leaching water—will give in the beginning of the leaching with the solvent a precipitate of silver with some lead; afterwards, however, the silver diminishes, so that the precipitate of lead finally appears free of silver. Besides the sulphate of lead, sub-chlorides and oxy-chlorides are formed during the roasting which are not soluble in water, but are dissolved by the hyposulphite of lime; for this reason always some base metals will be found in the precipitate.

In case rebellious ores are treated, and hot water is used for the extraction of base chlorides, a better silver is obtained if the ore is cooled down by cold water before the cold and diluted solvent is applied. Purer ores may be treated with a warm solution of the solvent.

When examining the tailings as to the amount of silver left therein, it must be remembered that, after leaching out a quantity of metals by water and the solvent, the ore lost a considerable part of its original weight, and that consequently one-half ounce of such tailings taken into assay will always give a larger silver button than there ought to be. A true assay of leached tailings is made if half an ounce of the same ore is leached on a filter with hot water and hyposulphite of lime, in the same way as the ore on a large scale, washed with water, dried and weighed. The weight found after leaching must be taken for half an ounce in assaying the tailings.

The residue, or tailings in the leaching box, must be removed now as valueless. The sides of the leaching boxes are from eighteen to twenty-four inches above the bottom, and being from six to eight feet square in the clear, the removing of the tailings by means of chloride is easily effected. The men must be careful not to dig too deep, otherwise the filter will be injured. It is quite proper to fix wooden staves, as long as the box requires, on top of the filter. These staves are one inch wide and one-half of an inch thick, and are placed from four to five inches apart, so as to protect the canvas or filter against the shovel. In Fig. 11 the staves are laid upon the canvas.

The leaching boxes or tubs may be arranged so that, being tipped over, the whole charge falls out at once. In this case the filter must be made in a different way from that described above. Rocks are not serviceable here. On the false bottom, *a*, of Fig. 10, a layer of thin, leafless switches is placed, and on this another one crosswise, then covered with a piece of canvas, and secured with some staves to prevent the falling out of the whole filter when turned over.

Precipitation of the Silver.

62. The liquid of the second leaching is conveyed through a trough or india rubber hose to the

Fig. 12.

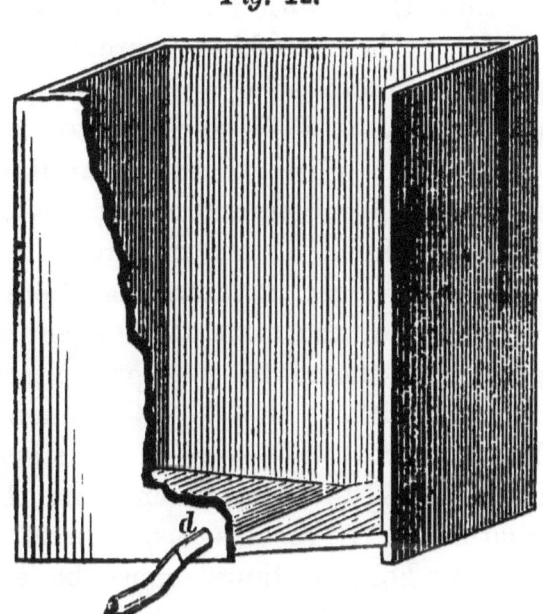

precipitating tanks, of which three or four are employed. If tubs are used, which for this purpose are the best, they are from three to four feet in diameter, and four feet high. The tanks or

boxes have a rectangular shape of about the same capacity, the bottom being inclined toward the middle, as shown by Fig. 12. The hyposulphite of lime, as it comes from the leaching tanks, is conducted into these until they are more than two-thirds full. . The trough or hose is then changed to discharge the liquid into the next precipitating tub, while the precipitation of the first commences.

The liquid used for precipitating the silver is sulphide of calcium (§ 69). It is poured in until all the silver is supposed to be precipitated, and at the same time the solution is stirred vigorously. Treating always the same kind of ore, the required quantity of the precipitating agent is soon learned. The black precipitate sinks to the bottom, and the workman now dips a little of the clear liquid out in a glass tube or tumbler, and adds a few drops of the sulphide of lime. If a dark precipitate or a dark color is produced, it shows that there is still silver in the liquid, and more of the agent must be added ; but if on the contrary no precipitate is observed, there is either enough or too much of the sulphide. To prove this, some of the silver-holding liquid is added to a test, taken from the tank under treatment. If in this case a precipitate is formed, silver-holding liquid must be carefully added to the tank until no reaction is produced. This work, delicate as it seems, is easily learned by the workmen. If a little silver should be left in the liquid, it is not injurious, neither is the silver to be considered as lost, because the same liquid is used over again ; but a small excess of the sulphide of calcium would cause a loss in silver, as it precipitates sulphide of silver in the leaching tank in the mass of ore, which is not dissolved again. The precipitation is performed in a short time, requiring about fifteen minutes for each tank. The stirring must be executed with vigor. Wooden grates fixed to a vertical stem will answer the purpose.

The clear solution above the settled precipitate is pumped or elevated to the reservoir, whence it was conveyed on the ore. It is now ready to be used again. The sulphide of calcium having performed its duty in precipitating the silver, is turned into hyposulphite of lime, thus replacing all of the solvent.

To prevent small floating particles of silver from being elevated with the liquid, it is well to allow sufficient time for the precipitated silver to settle. For this reason it is better to have more precipitating tanks or tubs. It is not necessary to remove the silver after each precipitation. The clear liquid can be drawn off, by means of a syphon, from all the precipitating tubs into a general receiver, whence it may be .pumped up. After the solvent has been removed, the precipitated silver can be drawn off through the pipe, *d*, Fig. 12, directly into canvas bags.

Treatment of the Precipitated Silver.

63. The black precipitate of sulphide of silver is conveyed directly into filters made of canvas, either in the shape of pointed bags, like those used for amalgam, or in the shape of common bags. As soon as all the liquid runs out, pure water (if possible, warm) is poured on the silver, and this repeated several times till no taste is observed in the filtering water. The precipitate, while still in the bags, is placed beneath a screw press and the fluid pressed out as completely as possible. The black silver cakes are then taken out and dried in a warm room or in a drying oven. For the purpose of burning off the sulphur, the dried sulphide is introduced into a muffle or other calcining furnace, and heated till the sulphur commences to burn with its known blue flame. When this disappears the heating must continue at a dark red heat for one or two hours. By this operation the cakes are reduced almost entirely to metallic silver, generally covered with threads of silver; sometimes an intense green color is assumed by pieces remaining in the furnaces over night.

The burned cakes are now prepared for smelting in crucibles. They are placed in black lead crucibles, according to the size, up to three hundred pounds, and fused. All the sulphur was not driven out by the preceding operation. The remaining part must be removed by placing metallic iron (§ 5, *d*) in the fused metal; thereby iron matt is formed, which rises to the surface and is skimmed off. The surface of the silver is then cleaned by adding some bone ash and borax, or borax alone, which is also skimmed off and the silver dipped out or poured out into moulds. According to the careful treatment in the roasting process, and the nature of the ore, the silver will be from 800 to 950 fine.

Mr. O. Hofmann, in need of sulphur for the production of sulphide of calcium, used to calcine the dried sulphide of silver in iron retorts. In this way he obtained a large proportion of sulphur as a fine sublimate. This could be done also in a proper muffle furnace, so arranged that after all obtainable sulphur had sublimated in a receiver this could be removed and the calcination continued under access of air.

Precipitation of Copper contained in the Ore and of a small amount of Silver leached out with the Copper.

64. Having refractory ore under treatment, it is generally the case that copper is also found in it. While roasting, the presence of copper is favorable for the chlorination of the silver, but copper ores require some more salt, especially if it is intended to save the copper also. The more chloride of copper formed, the more will be found in the solution while leaching it with hot water. In order to convert all the copper into a chloride, it would take at least one and a half pounds of salt to each pound of copper; and considering other base metals, lime, etc., all of which absorb chlorine, while a considerable part escapes useless, the above

quantity has to be doubled. For this reason no
special attention can be paid to the copper; only
that part of it can be extracted which is converted
into a chloride during roasting under the usual
circumstances. The chloride of copper transfers a
part of its chlorine to the silver and other metals
(§ 23), and is reduced thereby to a sub-chloride;
if there is sufficient salt in the furnace it is raised
again to a chloride. This sub-chloride ($Cu^2 Cl$) is
not soluble in water; it remains in the ore during
leaching.

65. The different chlorides, being removed in
the first leaching (§ 61), are principally those of
copper, iron, lead, antimony and zinc, besides some
undecomposed salt. The first quantity of hot
water introduced into the leaching box is, of
course, most saturated with the named salts, and
they have the property of dissolving, also, some
chloride of silver. The dissolved silver precipitates again as soon as it becomes diluted with more
water. There is, therefore, no difficulty in regaining the silver which is thus leached out. The
amount of silver carried out by the leaching water
varies from 0.5 to three per cent. Not only the
chloride of silver, but also those of lead and antimony, are precipitated by dilution with water.
There are two ways of regaining this silver.

Mr. O. Hofmann adopted an ingenious plan for
this purpose, by conveying the hot water, under a
slight pressure from below, through the pipe, e,

(§ 61, Fig. 10), by attaching to it a rubber hose. The water rises through the ore from e up to d, and as soon as it reaches within two inches of the brim of the box, the hose is removed from e, and the water admitted through d. The concentrated solution, containing dissolved silver, is now above the ore, and being diluted with water from d, lets the chloride of silver fall as a precipitate on and throughout the ore.

The other plan is the precipitation of the silver, together with the chlorides of lead and antimony, outside of the leaching box. This mode is preferable to the former when a great deal of lead and antimony is in the ore ; for if precipitated in the box, a great part of it will be dissolved by the hyposulphite of lime and then precipitated as sulphides with the silver, making this impure and consuming much of the precipitating agent. As soon as the chlorides flow into the trough, f, below e, into which several leaching boxes discharge their fluids in different degrees of dilution, the gradual precipitation commences. The precipitate is white and adheres to the trough through the whole length of it. These chlorides are the richest, and contained, at Flint, Idaho, 9 per cent. of silver; the balance was principally lead and antimony. The precipitate deposits on all bodies offering a surface. For this purpose a box must be constructed, as represented in Fig. 13, which shows the top view or ground plan. The sides, a, of a wooden box, six feet by six, or ten by six, are six inches high, the front, *b'*, and partitions, *b*, four inches high, leaving a space of six inches between them. These spaces are filled with shavings representing an immense amount of surface for the chlorides to de-

Fig. 13.

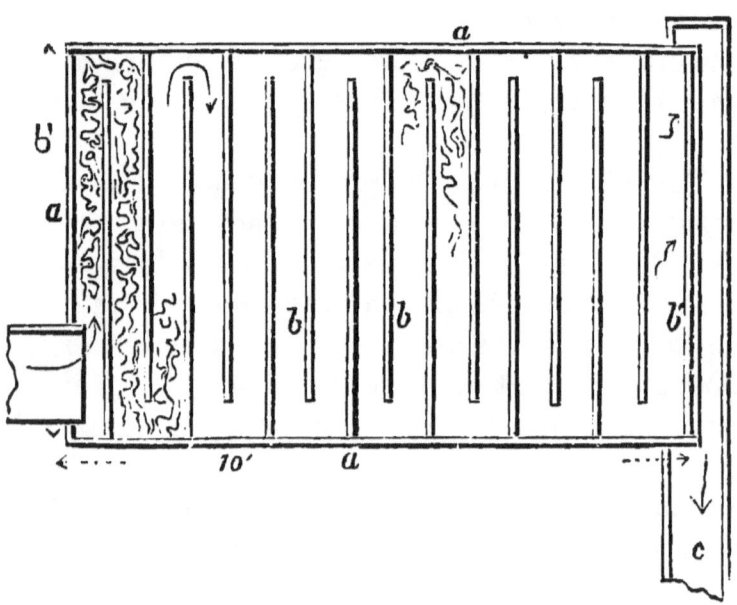

posit on. The fluid entering the trough, *c*, contains now a purified copper solution. Chloride of iron is also with the copper in solution, but does not prevent the copper from precipitating.

66. The white precipitate, when accumulated, is taken out, placed in filtering bags, with or without the shavings, and washed with clear cold water, in order to get rid of the copper solution. The silver can be extracted in two ways : The simplest mode is the application of hyposulphite of lime.

The sediment is taken out from the filtering bags and charged, while wet, into a filtering box of a proper size, arranged like Fig. 11, § 61. The hyposulphite of lime, in a cold condition, is poured over it and managed as with the ore with the second leaching, § 61. The silver-holding fluid may be conveyed into the precipitating box, Fig. 12, § 62, and treated with the solution from the ore. The liquid from the bags is examined from time to time with sulphide of calcium. In the beginning the precipitate appears dark, being mostly silver; but when it is perceived that the precipitate assumes a light yellow color, too much of lead, zinc and antimony is being carried out, and the leaching must be stopped. The residue in the filter box contains still some silver.

The other mode of extraction is more perfect, but also more expensive and more troublesome. After the copper has been washed off, the contents of the bags are taken out and dried. It is then introduced into large crucibles and smelted with an addition of soda-ash. The reduced metal, if some lead occurs in the ore, must be separated by means of cupellation, resulting in clean silver and litharge.

67. The chloride of copper running from the box, Fig. 13, is led into a reservoir in which old iron is suspended. The copper precipitates in a metallic state on the iron, and about eighty-eight parts of iron go into the solution in place of one hundred parts of copper; consequently, as each one hundred pounds of pure copper require eighty-eight pounds of iron, the calculation as to the necessary amount of iron could be made easily if it were not for some other chlorides which may still be in solution, and which also require iron for precipitation. Wrought iron is preferable to cast iron, and gray cast iron is better than white; but all these sorts precipitate the copper, and it depends to a great extent on the price as to what kind of iron is chosen.

The most effective precipitant is the iron-sponge or finely divided iron, obtained by heating pulverized iron ore or roasted iron pyrites with charcoal powder in a proper reverberatory furnace, or in iron pipes or cylinders without admitting air. By these means the iron oxide is reduced to metallic iron, which precipitates the copper in a few minutes. Using old iron, the precipitation will be effected much more quickly than in tanks if an arrangement is made like Fig 13, putting the old iron in the place of the shavings. This would be a continuation of Fig. 13, but the box must be three or four times as long before it reaches the tank or reservoir.

To find out whether there is yet copper in the liquid, the best test is to take some drops on a piece of platinum, and to place a small, clean piece of zinc on it. The copper immediately appears of a bright red color. But for practical use, a clean piece of iron dipped into the liquid will also show a red coating if there is enough copper in it to make it remunerative to continue the precipitation. The water contains now principally chloride of iron, and is discharged. If by some cheap means the water could be evaporated, the remaining chloride of iron could be used in roasting ores without salt.

Where old iron commands a high price, the copper can be precipitated with a brine of ashes or of lime ; but in this case the iron also falls with the copper. The brine for this reason cannot be advantageously adopted where a great deal of iron is in the ore, or the roasting must be directed so as to decompose the chloride of iron.

Quality of Ores fit for the Solving Process.

68. There is no process so suitable for all kinds of ores as the solving process. Generally considered, all silver ores can be treated by the solving process which are subjected to the pan amalgamation after roasting ; but in many instances—especially with the rebellious ores—a better result is obtained by this than by working in pans. The great advantage of this process is cheapness. Roasting of course is indispensable except with chloride ores ; but neither pans and the required power, nor quicksilver, are used, and for this reason less capital is necessary to put up reduction works. All the cupreous silver ores of Cerro Gordo, Yellow Pine, Montgomery, and of the other new silver districts, can be treated to great advantage by the solving process.

6

Two objections have to be considered. First, there is more water required than for pan amalgamation—at least this is the case with rebellious ores; but the quantity of water depends on the quantity of base metals in the ore, and also on the arrangement of the leaching boxes. One box, containing one ton of ore, requiring three hours leaching, may consume 250 gallons of water; three boxes of the same size would take three times as much water if placed on the same level; but by arranging the boxes in a less favorable position, one above the other, as shown in Fig. 14, only one-half of the quantity of water is needed. It takes more time to leach three boxes together than a single one. Leaching the ore with hyposulphite of lime, the supply of this in the first box must be stopped if no more silver comes out, and the solution carried, by means of hose, to the second,

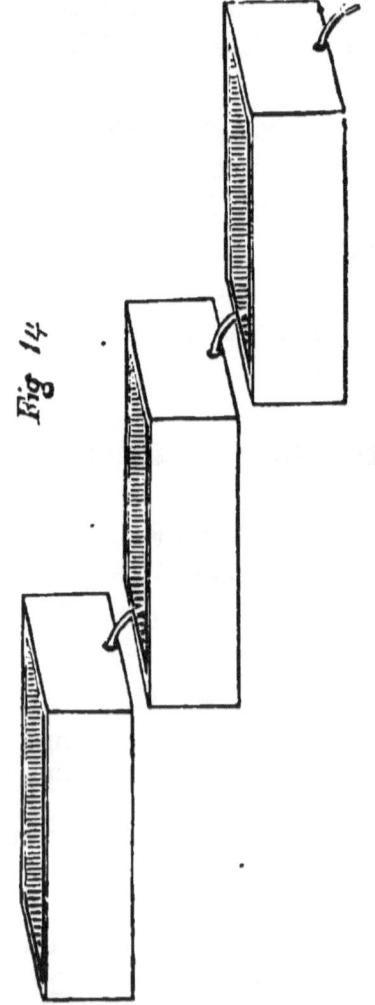

Fig 14

and then to the third box, if it should be necessary; so that while the second box is yet under
leaching, the first can be discharged and a new
charge introduced. All three boxes should receive
clean water at the start. This arrangement should
be adopted only when rendered necessary by the'
scarcity of water. To have all leaching boxes on
a level is preferable.

The other objection is confined to a certain class
of ores containing clay and lime. If pulverized,
so much fine pulp will be produced that the leaching is impossible. It is not advisable to crush the
ore coarser than will allow of its passing through a
sieve of forty holes to the inch, in some cases,
perhaps, through thirty-five holes; and if with
such crushing a fine clay pulp is produced, the ore
is unfit for all leaching processes, unless wet crushing is adopted, in order to separate the slime from
the sand, as Mr. O. Hofmann was compelled to
arrange in Trinidad, Sonora. In this case, a separate drying for the purpose of roasting is not
necessary if long furnaces are in use. It is not
unlikely that for similar ore and pan tailings an
agitating filtering box could be constructed which
would render the leaching possible.

Sulphide of Calcium.

69. Sulphide of calcium for the precipitation
of silver is preferable to the sulphide of sodium,
principally for the reason that its manufacture is

cheaper and more easy, but also on account of the quality of the precipitated silver, which is easier to wash, to press and to desulphurize. The sulphide of calcium is easily obtained and manufactured on the ground where the mill is situated. The articles required for this purpose are brimstone (worth about four cents per pound) and burned lime. The sulphide is formed only from caustic lime, consequently more is obtained from fresh burned lime. Of this a certain quantity is charged into an iron kettle, water added, and then the pulverized sulphur. The proportion of sulphur and lime depends on the quality of the latter. The purest quality of lime from Santa Cruz, Cal., for instance, takes one pound of sulphur to 1.33 of lime. Of poorer qualities of lime it is better to take three pounds to one of sulphur and about ten parts of water. It is kept boiling for two or three hours, stirred with wooden shovels, and then discharged into a filtering box, prepared like Fig. 10, § 61. The clear, dark, yellow-red sulphide of calcium comes out from below the filter, and can be kept in iron vessels. The liquid ought to be between 5° and 6° Beaumé. The residue is washed with water, whereby a diluted fluid is obtained, which is used with the lime of the next charge. Mr. E. Smyth, in La Dura, Mexico, treats the lime and sulphur with steam. This has the advantage of dispensing with the stirring, and may be performed also in wooden vessels. The steam replaces the fire and has no chemical influence on the quality of the sulphide ; the precipitating capacity of the latter with reference to the volume depends only on the concentration which is expressed by the degrees of Beaumé's hydrometer.

One pound of lime (Santa Cruz quality) gives sufficient sulphide of calcium to precipitate one and a half pounds of silver.

Hyposulphite of Lime.

70. The hyposulphite of lime as a solvent of chloride of silver has a great advantage over a hot solution of salt. It can be applied diluted and cold, and dissolves a great deal more of the chloride than does the salt, of which nearly sixty-eight pounds are required to dissolve one pound of chloride of silver, while only two pounds of the hyposulphite are needed to dissolve the same amount of the chloride.

The hyposulphite of lime is produced by conveying sulphurous acid into sulphide of calcium till it appears entirely colorless. It is also formed if from a concentrated brine, obtained from lixiviating roasted ore with hot water, all chlorides are precipitated by sulphide of calcium. After the precipitated sulphides have settled, the clear fluid can be used to dissolve the chloride of silver. The simplest way, however, is to buy hyposulphite of soda, and to commence the second leaching (§ 61) therewith, precipitating with sulphide of calcium.

Patera Process.

71. The most delicate operation in Patera's process is the preparation by roasting, as described in § 33. The chloride of silver formed during the roasting is dissolved by a cold solution of hyposulphite of soda, after all soluble base metals have been first leached out with hot water (§ 61). Two parts of the hyposulphite of soda dissolve one part of chloride of silver, forming a soluble double salt. The tubs in which the ore is lixiviated with the hyposulphite of soda are small, receiving only 200 pounds of roasted ore. The extraction of silver is performed in the same way as described under § 61.

Kiss Process.

72. This process extracts silver and gold together. Roasting the ore, as explained in § 38, the gold is transformed into such a state as to render it insoluble in water. After roasting, the ore is placed in filtering tubs and washed with water to remove the base metals. After this a solution of hyposulphite of lime is conveyed on the ore, by which the gold and silver chlorides are dissolved and carried off into precipitating tubs. As soon as the sulphide of calcium is introduced, the gold and silver are precipitated as sulphides. The precipitation of both metals in a metallic condition is not admissible, for the reason that the hyposulphite of lime is decomposed if metallic copper is employed for precipitation. The results of · Kiss's methods, practiced in Hungary, were not altogether satisfactory.

Patera and Rœszner Processes.

73. The object of these processes is, like that of the preceding, the extraction of silver and gold together. The ore is first subjected to a chloridizing roasting, by which the silver is converted into a chloride, while the gold remains mostly in metallic condition. The leaching liquid is prepared by conveying chlorine gas through a cold concentrated solution of salt to saturation. This chloridized solution dissolves silver, gold and copper at the same time. The roasted ore is charged into tubs with false bottoms, and the cold solution of salt and chlorine introduced. Silver ores treated after this method in Hungary gave 98.94 per cent. of the silver, all the copper and nearly all the gold. An experiment on five tons of ore gave a clear profit of seventy-five florins, compared with the amalgamation.

Röszner roasts the ore with salt, extracts a part of the silver by Augustin's method with a hot solution of salt, and treats the residue alternately with a solution of salt and chlorine, and with a hot concentrated salt solution for the extraction of gold and the remainder of the silver.

· Kustel & Hofmann Process.

74. Auriferous silver ores are roasted, as described under § 35. They are then subjected to a
process differing from the preceding ones in obtaining separately the copper, gold and silver.
After roasting, the ore must be moistened on the
floor, by conveying water through a spout, and
mixed with shovels, so as to get it uniformly moist,
but not so wet as to interfere with sifting, which,
however, is not always necessary. The ore is then
put into tubs or boxes, as represented by Fig. 10
or 11, § 61, but provided with covers, which can
be easily screwed air-tight on the rim of the box,
having india rubber between the box and the
covers. The ore should not reach the brim of the
box, but leave a space of four or five inches at
least above the ore, as a chamber for surplus chlorine. When the box has been filled, the cover is
screwed or otherwise fastened on the box, and the
chlorine gas admitted (§ 11).

The chlorine gas is generated in a leaden generator of the construction shown in Fig. 15, which
represents a cross section. It is a circular tub,
with an outer ring, a, six inches deep, for the reception of the cover, b. There is a similar ring, c,
on the top of the cover, which receives the collar
fastened to the leaden stirrer, d. There is also a
short leaden pipe, e, bent in the shape of an s,
through which the sulphuric acid is introduced.

Another lead pipe, f, conveys the chlorine out of the generator. The vessel is uncovered, and for a charge of three tons of roasted ore the following ingredients are introduced: Thirty pounds of peroxide of manganese (pulverized); thirty to forty pounds of common salt, according to quality; seventy-five pounds of sulphuric acid, of sixty-six degrees; and forty-five pounds of water. Salt, manganese and water are introduced first, and the generator covered. The two rings, a and c, are filled with water, in order to close the generator air tight. The sulphuric acid is now charged through the funnel, e, in small amounts; twenty to twenty-five pounds are sufficient to evolve the chlorine and the required heat. When the evolution of chlorine becomes weaker, twenty pounds more of acid are administered, and after some time the rest of the seventy-five pounds. It will be

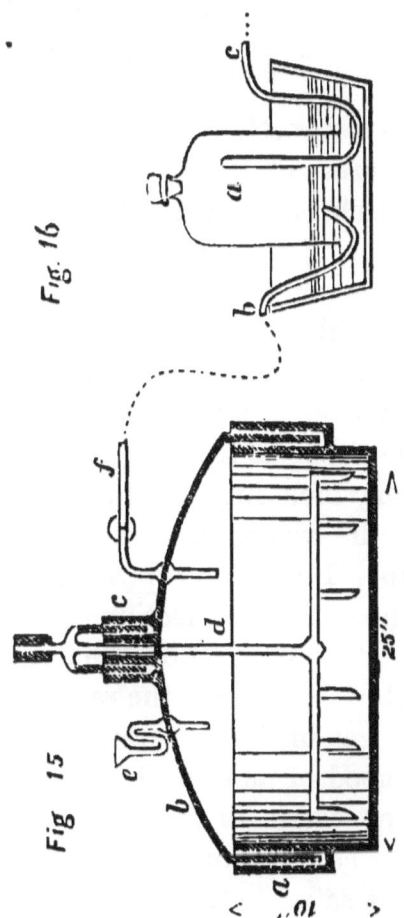

6*

necessary now to build a fire beneath the generator, which is placed on tiles that the heat may not injure the leaden bottom, which is made of sixteen-pound sheet lead, while the sides are of eight-pound.

The chlorine is not led directly to the ore, but through a purifying apparatus, as represented in Fig. 16. An ordinary wash basin or a similar vessel receives the lead pipes (three-quarter inch). One of these, *b*, conveys the chlorine from the generator ; the other, *c*, leads the gas to the ore-box. Both ends are covered with a bottle, *a*, the bottom of which is cut off (by means of a string). Clean water is poured in the basin so as to cover the end of the pipe, *b*, about one inch high. The gas is forced to pass through the water, by which the muriatic acid is taken up. Both lead pipes can be provided with rubber hose for connections with generator and ore-box. The passing of the chlorine through the water shows distinctly the rapidity of the evolution of the gas in the generator, and indicates when there is need of more acid or of the application of heat. The water in the basin absorbs more chlorine when cold—about $2\frac{1}{2}$ per cent. of its volume—before it is saturated. A continual stream of water is therefore improper.

The purified chlorine is now conducted through the pipe, *e*, of Fig. 10 or 11, § 61, to which the rubber hose or lead pipe from the purifying apparatus (Fig. 16) is attached. The chlorine goes through the whole mass of the ore, driving out the lighter air through a hole in the cover till the gas itself comes out at the same hole. A glass rod is dipped from time to time into ammonia, and held before the hole. As soon as white fumes appear on the rod, it is proof that the box is filled with chlorine. The hole is now closed and the fire below the generator removed. In this condition the ore remains for twelve or fifteen hours. The whole arrangement must be examined with ammonia, to see that there is no loss of chlorine. According to the amount and quality of the gold, it may be necessary to allow the chlorine to operate for eighteen to twenty hours.

Lixiviation. It is generally the case, and ought to be so, that only a part of the chlorine is consumed, while the rest is unchanged. For the sake of economy and for sanitary considerations, the free escape of the surplus gas should not be allowed ; but this should be utilized for the same purpose— that is, for chlorination. This gas is easily transferred to another box prepared with moistened ore. Mr. O. Hofmann inserted a rubber pipe into the hole of the cover of the ore-box, and joined the other end with a pipe, *e,* of another box, as in Fig. 10. Through the pipe, *e,* of the already chloridized ore, he admitted the water, which, entering the box, displaced the chlorine and forced it into the other vat. The water, however, dissolves a part of the chlorine, and for this reason it is better to convey the surplus gas over by suction, which is easily effected by an air-tight tub in which a

vacuum is created by the discharge of water. This
way is the more advisable, as a delay, by which the
leaching water is longer in the box, is injurious to
the gold. When the gas is removed, hot water is
introduced, and the leach, containing gold, copper, etc., led into the precipitating box. An addition of dissolved sulphate of iron will precipitate
the gold in metallic condition. The sulphate of
iron is either procured as an article of commerce,
or prepared by throwing old iron into diluted sulphuric acid.

After several hours (from eight to twelve) the
clear solution is drawn off from the precipitated
gold, and if the copper is to be extracted, conveyed
to other vats and treated as described in § 67. The
liquid, coming out through the filter of Fig. 10,
must be examined in the beginning with a clear solution of sulphate of iron, and as long as a dark
color of precipitated gold is perceived, the leach is ·
allowed to run into the gold precipitating tub ; if
there is no precipitate observed, the leach is di-
. rected into the copper vats. The liquid is now examined with sulphide of sodium or of calcium, and
proceeded with exactly as shown in § 61, '' Second
Leaching.''

Augustin Process.

75. This process is not in use at present for
silver ores, but for products of smelting. By this
method the chloride of silver, which is formed by

way of roasting (§ 30–31), is dissolved in a hot solution of salt, and precipitated by metallic copper.
One part of chloride of silver requires sixty-eight
parts of salt, to be dissolved.

*Extraction of the Silver from Copper Matt and
Black Copper.* The principal aim with these materials is the oxidation of the copper as perfectly as
possible, and then the chlorination of the silver.
There are wooden leaching tubs of a small size—
two feet eight inches in diameter, and nearly four
feet high—fixed on wheels and arranged in one
row. Into these tubs, which have false bottoms,
the roasted stuff is introduced—about 800 pounds
in each. Ores containing different kinds of earths
cannot be lixiviated at a depth of over three feet;
the metal oxides, however, allow the water to pass
freely. This is also the case with roasted, concentrated, or pure sulphurets. Hot solution of salt is
now allowed to flow through a trough in each tub.
The salt penetrates the powder, dissolves the chloride of silver, and carries it through the filter at
the bottom of the tubs, flows off to a reservoir, and
from here, after the particles which may escape
through the filter have settled, into a series of vessels one above the other. These are provided
with double bottoms. The two uppermost rows
contain cement copper, six inches deep ; the low-
est, metallic iron.

The fluid deposits its silver principally in the
first tub, dissolving at the same time an equivalent
amount of copper. Some silver which escapes pre-

cipitation falls with the cupreous fluid into the next
tub below, where the rest of the silver is taken up
by the copper.   In the third vessel the copper is
precipitated by the iron.   The brine, freed from
silver and copper, is pumped up into the reservoir,
heated and used again.   The cement copper obtained in the last tub is placed back in the upper.
two.   The brine circulates in the tubs until a
bright copper plate is not coated with silver when
held in the fluid from the leaching tubs.   The
residue, which is mostly copper-oxide, is removed,.
and an average sample taken and assayed.   If it
should contain over eight ounces per ton, it must
be roasted over and again lixiviated.

The precipitated silver is taken out once a week
and treated with muriatic acid for the purpose of
dissolving the copper particles which remained
with the silver.   After this it is washed with water
till all traces of the acid disappear, then pressed
into balls, dried and melted.

Ziervogel Process.

76.   Like the preceding, Ziervogel's extraction
of silver is not applied to silver ores, but only to
copper matt.   The roasting (§ 43) is very delicate,
and it is more difficult to obtain a satisfactory result with silver ores than by a chloridizing roasting.   The silver in this process is converted into a
sulphate, which is soluble in water, thus dispensing with the expensive salt brine.   The pulverized

and properly roasted copper matt is charged into leaching tubs, 500 pounds in each, and hot water admitted. As soon as the water begins to flow out, the hot water is made a little acid by admixture of some sulphuric acid. The lixiviation continues until a sample of the fluid remains clear if a solution of salt is added. The silver-holding brine is conveyed into a large reservoir, thirty feet long, where it clears of impurities, which accidentally come out of the leaching tubs, and falls from this reservoir through a series of cocks into the precipitating tubs. On the false bottom is a layer of cement copper, and upon this fifteen to twenty copper bars of 250 pounds weight. Each is fourteen inches long, five inches wide and one inch thick. The liquid loses most of its silver in these tubs, and flows then through a trough fifteen inches wide, lined with sheet lead and having a layer of copper pieces on the bottom, into five vats filled with copper, where the balance of the silver is deposited.

The desilverized brine comes now into a reservoir, whence it is pumped up into a large leaden pan and heated again by means of steam. Above this pan is a leaden vessel, out of which about thirty drops of somewhat diluted sulphuric acid drop into the liquid every minute. The acid prevents the separation of basic salts. The silver is taken out of the precipitating tubs every day. With it occur some copper and gypsum. The larger particles of copper are separated by washing, exposed for six

or seven days to leaching with diluted sulphuric
acid, and finally washed with hot water. The silver is from 860 to 870 fine. After drying, it is re-
fined in a reverberatory furnace.

Once a year the brine is brought into contact
with iron, in order to precipitate the copper. The
purer part of the cement copper is used for the
silver precipitation, and the finer part is delivered
for smelting.

The Leaching Process.*

77. Under this name is understood a preparation of the ore applicable for the pan amalgamation.
Its description, therefore, does not belong here
strictly, but the leaching itself has so close a connection with the preceding manipulations that this
part alone may be described without mentioning
the further treatment by amalgamation.

It is a known fact that, in treating refractory ore in
pans by amalgamation, of course by way of roasting, some very annoying things are encountered,
and amongst them principally, the great loss of
quicksilver, amounting sometimes up to ten or
twelve pounds per ton of ore ; the rapid destruction
of pans, which compelled many mills to use
wooden sides fixed to the iron-pan bottom, a meas-
ure which saves the pans at the expense of quicksilver; and the very base bullion which results from

---

*The Leaching Process is patented, as an application for pan and
barrel amalgamation, by G. Kustel.

such a treatment. In some instances it happens that a great deal, sometimes over fifty per cent., of iron goes into the amalgam, rendering the continuation of the amalgamation impossible. The result of the amalgamation of base metals is always a certain loss of silver, which would have amalgamated if the base metals were out of the way. It happened very often in Nevada that $90 to $100-ore was purchased for the purpose of amalgamating it in pans; but a few tons proved that amalgamation had to be given up. Such ore is now considered suitable only for smelting.

At a very trifling expense all these difficulties can be avoided and the amalgamation turned into a perfect success ; for instance, the amalgamation of the silver ores at Flint, Idaho, (§32) turned out such base amalgam that further working proved to be ruinous. The introduction of the leaching process, however, resulted in a most favorable amalgamation. It is only to be regretted that after working several hundred tons, the mine refused to provide the mill with ore, perhaps on account of not having been sufficiently opened. The leaching for the pan amalgamation is most important and at the same time cheap ; all the expense is reduced to that of obtaining hot water. This process is not only important for silver ores containing base metals, but also for gold ores which by their nature require roasting. This refers principally to auriferous copper ores, as the amalgamation of gold is very much obstructed by the presence of copper salts.

It is a matter of surprise how so simple a remedy could have been overlooked while fighting with the obstructions, caused by rebellious ores, during the amalgamation.   If there is soluble chloride of sil-ver in the roasted ore, and besides this, soluble chlorides of copper, lead, antimony and zinc, it is a matter of course that all will be decomposed and amalgamated.   All take part in consuming and parting the quicksilver, and in destroying the pan, hindering at the same time the easy amalgamation of the silver and gold.   Why, then, not put all these obstructive metals out of the way and give the silver a better chance to amalgamate ?   The base metal chlorides are soluble in water, the chloride of silver is not.   It is therefore a most simple ma-nipulation to dissolve those salts in water and to remove them from the ore before amalgamation, by the leaching process.   As soon as this is done the ore is divested of its rebellious nature and it be-haves in pans like the best ore.

The process of leaching is described § 61 a.

IV.   EXTRACTION OF GOLD.

78.   The extraction of gold without the use of quicksilver is limited mostly to those ores in which the gold is not free in a metallic condition, but combined with sulphur or arsenic in the respective pyrites.

There is only one body with which the gold must be combined before subjected to further treatment, and this body is chlorine. The chlorination can be effected either during roasting, § 38, or after roasting, by contact with chlorine gas, § 74, 79, or finally by contact with chlorinated water, § 73. The usual way of chloridizing the gold is that by introducing the chlorine gas into the roasted ore, when cold. This mode is described fully in Kustel's work on " Concentration and Chlorination." No improvement of importance has been since introduced in · this process.

The Chlorination Process. (Plattner's.)

79. This process is based on the property of metallic gold of being changed into a soluble chloride of gold when in contact with chlorine gas. The chloride of gold can be dissolved in water, separated from the ore by lixiviation, and then precipitated in a metallic condition by a solution of sulphate of iron. There are several establishments in California, principally in Grass Valley, * where auriferous pyrites are treated by chlorination on a large scale. By way of chlorination, if properly executed, 90 to 95 per cent. of the fire assay can be extracted.

---

* The first idea of trying this process on sulphurets in California came from Mr. Ch. Von Beseler, who experimented on it with Mr. Deetken in 1858. Since then Mr. Deetken has been engaged in the process, up to the present day, superintending chlorination works in Grass Valley.

80. In order to be sure of a result on a large scale, it is an easy matter to make an experiment with twenty or thirty pounds of sulphurets or ore in the following way : The named quantity must be roasted first, and it is the most difficult task, requiring either a small furnace or a great deal of patience, especially when small charges are treated on a large piece of sheet iron, having a charcoal fire beneath. In either case the sulphur must be driven out perfectly, so that when in a glowing condition, no smell of sulphurous acid can be observed. When finished a scruple is taken for an assay, and the roasted stuff moistened with water, after the weight of the whole has been noted.

Fig. 17.

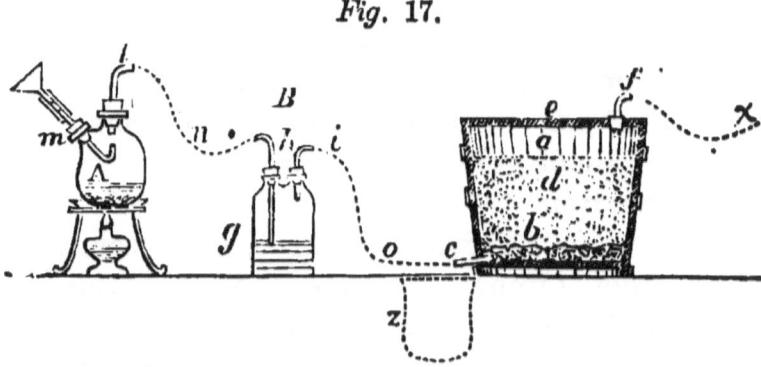

A common water bucket is then prepared to receive the moistened ore, which must not be too wet, but only moist enough to allow its being sifted. On the bottom of the bucket, *a*, Fig. 17, some clean rock or broken glass is placed about two inches deep, and covered with a piece of moistened canvas. A short glass pipe, *c*, two-eighths of an inch in diameter, is inserted close above the bottom.

The ore, d, is then introduced, filling up two-thirds or less of the space as loosely as possible, and covered with a wooden or iron cover and pasted all around with dough. The cover is provided with a short glass tube, like c, to which an india rubber tube, f, for carrying the gas out of the room is attached. Both glass tubes, c and f, must be likewise secured with dough.

The chlorine gas is generated in a glass vessel, A,*. There are two corks in it, each having a glass tube, as represented in the drawing. The cork, l, is removed and the vessel charged with 3 ounces of peroxide of manganese, 4 ounces of common salt, and $4\frac{1}{2}$ ounces of water—all of which are well mixed. The cork is inserted again and well secured with dough. Another vessel, B, provided with two necks, contains water as indicated by g; the glass tube, h, dips about one-half inch into the water. The corks are made air tight like the others in A. The whole apparatus is now joined together by rubber pipe, n and o, fitting tightly to the glass tubes. Having all thus prepared, $7\frac{1}{2}$ ounces of sulphuric acid are poured through the safety-tube, m, but only in small portions and at intervals. When the bubbling of the water at g, in the vessel B, is not lively enough, some more acid is introduced, and finally the temperature raised by an alcohol lamp. If all the joints have been luted carefully with dough, not the slightest incon-

* All materials necessary for such an apparatus can be procured from John Taylor, on Washington street, San Francisco.

venience will be met with.   The chlorine gas from
the generator, *A*, is forced through the water in *B*,
by this means washed from muriatic acid.   Through
the pipe, *o*, it enters the bucket and ascends slowly
till it reaches the cover, escaping then through the
rubber pipe, *F*, where it must be examined from
time to·time by dipping a glass rod into ammonia
and holding it to the end of the pipe, *x*, which
leads out of the room.   In contact with chlorine
the ammonia evolves white fumes, and chlorine can
be detected by these means wherever it may escape.
The gas is allowed to pass through the bucket as
long as chlorine is created.   In this condition, by
stopping up the pipe, *x*, if no more chlorine is
evolved the apparatus may stand until the next
day.   The cover is then removed, the pipe, *o*, taken
off, a clean glass or porcelain vessel, as indicated
by *z*,   placed below *c*,   and warm water carefully
poured over the ore till the bucket appears to be
full.   The solution which comes out at *c*, must be
examined at times in a small tumbler with a few
drops of a solution of sulphate of iron.   If the
clear solution remains unchanged, without becom-
ing darker, the lixiviation is finished.

To the solution in the vessel, *z*, a few drops of
muriatic acid and then sulphate of iron, or green
vitriol, (dissolved) is added and stirred with a glass
rod.   The whole is allowed to stand till all the
gold is precipitated and the liquid is perfectly
clear.   This is drawn off by means of a syphon,
for which the rubber pipe, *x*, can be used.   The

remaining fluid and the precipitated gold is gathered on a filter, washed with warm water and dried with the filter in a porcelain cup, above an alcohol lamp. The filter is burned either free or under a muffle, care being taken not to lose a particle of the filter ashes; mixed with some lead it is then cupelled and the gold button weighed. A comparison with the assay shows to what percentage the chlorination has proceeded.

Chlorination of Sulphurets and Arseniurets.

81. All the proceedings of the chlorination have been treated already in describing the process of silver to which we will here refer. The first operation is an oxidizing roasting, as explained in § 44. When roasted, the ore is moistened, charged into chlorinating vats—which are preferable to boxes for gold—and chloridized according to § 74. After several chlorinations have been performed and the gold has accumulated in the precipitating vat, the inside of which should be varnished with asphaltum varnish, the gold is taken out by means of a scoop, put into a clean porcelain dish or enameled vessel, filtered, washed first with diluted nitric acid and then with hot water, dried and melted with the addition of some borax.

Other methods of extracting gold without mercury are mentioned in § 38 and § 73.

TABLE OF CONTENTS.